SURV 11 - Creating the Future of Sustainable Fast Boats

Rotterdam, The Netherlands
21-22 June 2023

ISBN: 978-1-7138-9564-0

Printed from e-media with permission by:

Curran Associates, Inc.
57 Morehouse Lane
Red Hook, NY 12571

Some format issues inherent in the e-media version may also appear in this print version.

The Royal Institution of Naval Architects
8-9 Northumberland Street
London, WC2N 5DA
United Kingdom

Phone: 020 7235 4622
Fax: 020 7259 5912

publications@rina.org.uk

Additional copies of this publication are available from:

Curran Associates, Inc.
57 Morehouse Lane
Red Hook, NY 12571 USA
Phone: 845-758-0400
Fax: 845-758-2633
Email: curran@proceedings.com
Web: www.proceedings.com

160 YEARS OF HISTORY, KNOWLEDGE AND EXPERIENCE

SALUM ET CARINAE PIGNORA VITAE

THE ROYAL INSTITUTION OF NAVAL ARCHITECTS

SURV 11 - Creating the Future of Sustainable Fast Boats

21-22 June 2023, Rotterdam, The Netherlands

In Partnership With

DAMEN

RNLI Lifeboats

TOGETHER WE WILL SHAPE OUR FUTURE

THE ROYAL INSTITUTION
OF NAVAL ARCHITECTS

SURV 11 - Creating the Future of Sustainable Fast Boats

21-22 June 2023, Rotterdam, The Netherlands

THE ROYAL INSTITUTION OF NAVAL ARCHITECTS
8-9 Northumberland Street
London WC2N 5DA

Telephone: 020 7235 4622
Fax: 020 7259 5912

ISSN No: 2752-6569

CONTENTS

ADVANCES IN PERFORMANCE PREDICTION OF STEPPED HULLS

A Dashtimanesh, KTH Royal Institute of Technology, Sweden

SUMMARY

There has always been an interest in stepped hulls because of their high performance in comparison to other conventional high-speed planing hull. Designers and boat builders have been orienting toward the multi-stepped hulls solution to ensure efficiency in calm water and waves at high speeds. However, the complexities in step design and lack of reliable and fast computational methods have led to various problems in the design of this high-performance hulls. To this end, we have conducted a series of studies based on mathematical modelling, numerical simulations, and experimental tests to understand the physics of the stepped hull and the complexities of step design. Accordingly, we have developed computer programs that can be implemented for performance prediction of stepped hulls in both calm water and regular waves. This paper summarizes all our efforts during the past 5 years and shows how our developed tools can be used for future design.

1. BACKGROUND

The Baltic Sea is highly sensitive to the environmental impact of human activity. Pollution at sea affects all the countries in the Baltic Sea. The Baltic Sea is connected to the world's oceans only by narrow and shallow waters and it is estimated that full exchange of the Baltic Sea waters can take place over a period of 25-30 years [1]. Based on HELCOM report 2018 [1], chronic oil pollution remains a significant environmental impact in the Baltic Sea due to its long-term effects on the environments. COx, NOx and SOx, as well as other GHG (Greenhouse Gases) emissions will have a negative impact on the Baltic Sea environment by 2030. Therefore, various international and national organizations have made the rules and legislations to prevent and reduce sea pollutions. In 2018, IMO future emissions strategy calls for a reduction in total GHG emissions of at least 50% by 2050 compared with 2008 [2].

In the meanwhile, there is a growing amount of small, predominantly leisure, craft in the Baltic Sea Region, especially in the Scandinavian region. According to HELCOM [1] report 2018, Sweden, Denmark and Finland belong to the countries with the highest per capita boat ownership in the world. In 2015, the national small boat registry (Finnish Transport Safety Agency) lists over 195000 small boats powered by a combustion engine. Generally, about 3.5 million leisure boats are active in the Baltic Sea, mainly close to the coastal areas for recreational boating [3].

A detailed questionnaire survey conducted by the Swedish Transport Agency included qualitative information on the activities of 881000 leisure boats in Sweden, including fuel consumption. Numerous and increasing number of high-speed craft in the Baltic Sea leads to critical sea and coastal area pollutions, arising complexities in shipping traffic and safety as well as living of coastal inhabitants. It should be noted that not only nature protection demands a clean Baltic Sea but also fishing business, tourism and leisure boating are economic sectors that rely on a clean and unspoiled Baltic Sea.

It is worth noting that a small number of very big leisure vessels, over 100 GT, are only registered according to IMO regulations. Most of the national and international organizations exempt high-speed craft from their rules and legislations related to emission, pollution, energy efficiency and fuel consumption. In today's environmentally conscious world, maritime innovation is expected to reflect the increasing societal demands to minimise the impact of shipping on the environment [4]. Accordingly, current, and future ship designs are expected to comply with demanding environmental standards, requirements for new equipment and for sustainability within the context of lifecycle performance.

Accordingly, in 2021, the Green Small Craft (GSC) Interreg seed money project in the Baltic Sea Region [5] intended to reduce main negative impacts of high-speed craft cruising in the Baltic Sea by improvements in design and operation of future high-speed craft for a clean, safe and smart boating. The project considered different techniques, their advantages, and disadvantages as well as the areas in which they are most applicable. The project focused on three main solutions including [6]:

- Hydrodynamic design of high-speed craft for safety and energy efficiency: Reducing hull resistant by designing specific hull form such as stepped and tunneled hulls and increasing hull stability.

- Propulsion system design for better energy efficiency and lower noise emission: Developing conceptual designs and prototypes of green propulsion systems including electric propulsion systems.
- Automation and digitalization of high-speed craft operation for increasing safety, smart cruising and fuel efficiency.

Regarding the hydrodynamic design of high-speed craft, nowadays, the design of high-speed craft is strongly conditioned by two anti-synergetic needs: the first is the reduction of fuel consumption and improvement of the speed, and the second is the improvement of comfort on-board [7]. To reach a balance between these needs, it is necessary to study the effects of various parameters on hull resistance as well as seakeeping and manoeuvring motions.

Moreover, the need to design high-speed efficient boats has derived boat designers to seek many different types of hull forms. Stepped planing hulls have sparked up a tremendous interest in recent years, and few researchers have tried to provide an understanding of their performance in smooth/rough water [8, 9]. These vessels are used for a wide range of purposes (e.g. military & patrol, fishing, and leisure). However, there still is a non-trivial question for the designers to define the height, the position, and the shape of the step(s) since there is a lack of data and methods for examining the effects of these parameters on stepped hull performance in calm and waves as well as during manoeuvring motions.

Therefore, the authors have conducted a lot of studies during past years to develop fast and reliable mathematical models for performance prediction of stepped hulls in calm water and waves. Moreover, we have developed CFD (Computational Fluid Dynamics) models to study stepped hulls hydrodynamics. Recently, we have also developed a comprehensive stepped hull series by considering various step characteristics. The current article tries to address and summarize all these efforts to spread the developed knowledge within the relevant scientific and industrial community, more effectively.

After a brief definition of the problem, three different solutions including our developed mathematical models, CFD simulations, and our developed systematic series of stepped hulls in the towing tank will be discussed.

2. PROBLEM DEFINITION

Naval architects strive to design planing hulls using various transverse discontinuities in the bottom, better known as steps. Generally, three different forms of stepped hulls may be considered in a planing hull identified as (1) transverse step, (2) step pointed aft and (3) re-entrant V step as shown in Figure 1. In addition to step geometry, step height and longitudinal position of step are also important factors which influence the wetted area. Wetted area of each portion of two-stepped hull may be different. Generally, a two-stepped hull can be considered as three tandem planing surfaces (TPS) which have their own planing characteristics such as wetted length, wetted surface, local trim angle, local deadrise angle, Reynolds number, Froude number, etc. Therefore, it is vital to consider each planing surface, separately.

However, to compute planing characteristics of each individual planing surface, development of an appropriate model for prediction of flow separation from each step is essential. Unfortunately, there is no appropriate wake model as yet which can be used for stepped hulls. In fact, various combinations of depicted situations (a--f) in Figure 1(b) can occur at different Froude numbers. The hydrodynamic lift also distributes in three different parts. To recognise how lift distributes over hull, equilibrium equations need to be solved.

It is worth mentioning although there are many studies about calm water performance of stepped hulls, the effects of step characteristics (height, position, number and shape) are still unknown. Moreover, there are only a few research on seakeeping characteristics of stepped hulls. Although the wetted surface may dramatically change during turning, there is no study on manoeuvring motions of stepped hulls. Meanwhile, there has always been a demand from boat designers for fast and reliable methods for performance prediction of stepped hulls in calm water and waves and for prediction of their manoeuvring motions. Therefore, current author and his colleagues have tried to address these challenges during past 5 years by developing mathematical models, performing CFD simulations and conducting experimental campaigns.

(a)

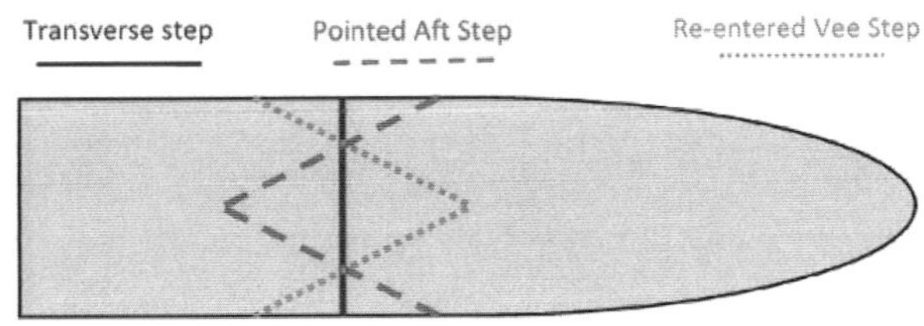

(b)

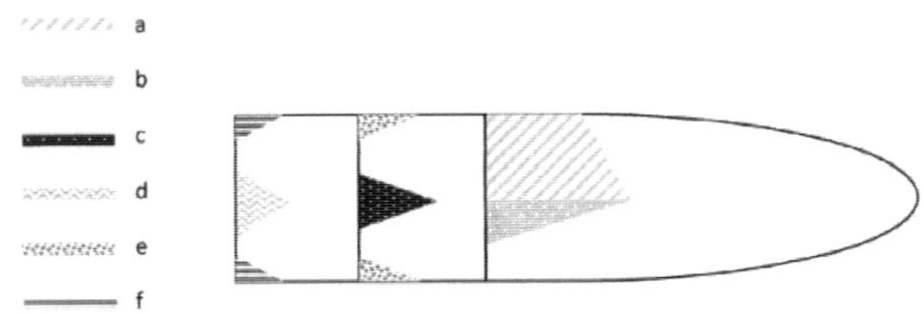

Figure 1: (a) Different forms of step, (b) Various possible wetting conditions for planing surfaces [10].

3. MATHEMATICAL MODELS

Most of boat designers are still using Savitsky 1964 method for performance prediction of stepped hulls in calm water, although Savitsky 1964 was only developed for simple prismatic planning hulls. To fill this gap, Svahn [11] extended Savitsky's method using wake theory for performance prediction of one-stepped hull. Danielsson and Stromquist [12] tried to develop Svahn's model for two-stepped hulls. However, they failed in implementation of wake theory of Savitsky and Morabito [13] for two-stepped hull because of range of applicability and other issues. Therefore, they assumed that flow separates from step and reattaches to after-body portion of hull by a linear trend. This assumption was reasonable in very high-speed situations. However, Danielsson and Stromquist [12] had some difficulties in definition of wetted length and Reynolds numbers. Moreover, there was a lack of complete and accurate validation for their model.

Subsequently, based on the above-mentioned developments, Dashtimanesh et al. [10] developed Savitsky 1964 for two-stepped hulls with transverse steps. Since there has been no direct measurement or empirical formula for wake profile beneath the two-stepped hulls, linear wake profile (LWP) was considered as a good assumption for flow separation from steps. They conducted an extensive validation study and showed that their method had a reasonable accuracy in comparison to experimental data. Based on their obtained results, it was observed that the presented model had insufficient accuracy in low speed range. Authors believed that mentioned inaccuracy was related to linear wake profile which was improper for low velocities. The accuracy of linear wake theory was also limited by the ship's displacement and step height. The linear wake theory was only usable for a particular range of step height. There was no sufficient experimental information to be able to recognise the relation between step height, displacement and LWP. Therefore, authors suggested that these relations must be quantified in future experiments by extracting and quantifying the flow pattern behind each step. Authors also noticed that the fluid flow spread out from step and chine in some situation which could lead to an added resistance known as spray resistance. In order to improve the mathematical model accuracy, authors proposed that other components of resistance such as spray resistance should be included in mathematical model. Another source of uncertainty was Savitsky's regression formula implemented in the mathematical model.

To address some of these challenges, a mathematical model based on the 2D+T theory was developed to predict the performance of two-stepped planing hulls in calm water [14]. It was attempted to develop a mathematical model without using regression formulas. It led to development of a computational model with no common limitations related to empirical models which have an individual range of applicability. Moreover, authors included spray resistance in their computations to improve the accuracy of the model. To evaluate the accuracy of the developed method, the obtained results were compared against experimental data and the previous empirical-based method [10]. The comparison suggested that the proposed 2D+T method predicted dynamic trim angle, wetted surface and resistance of double stepped boats with reasonable accuracy (Figure 2).

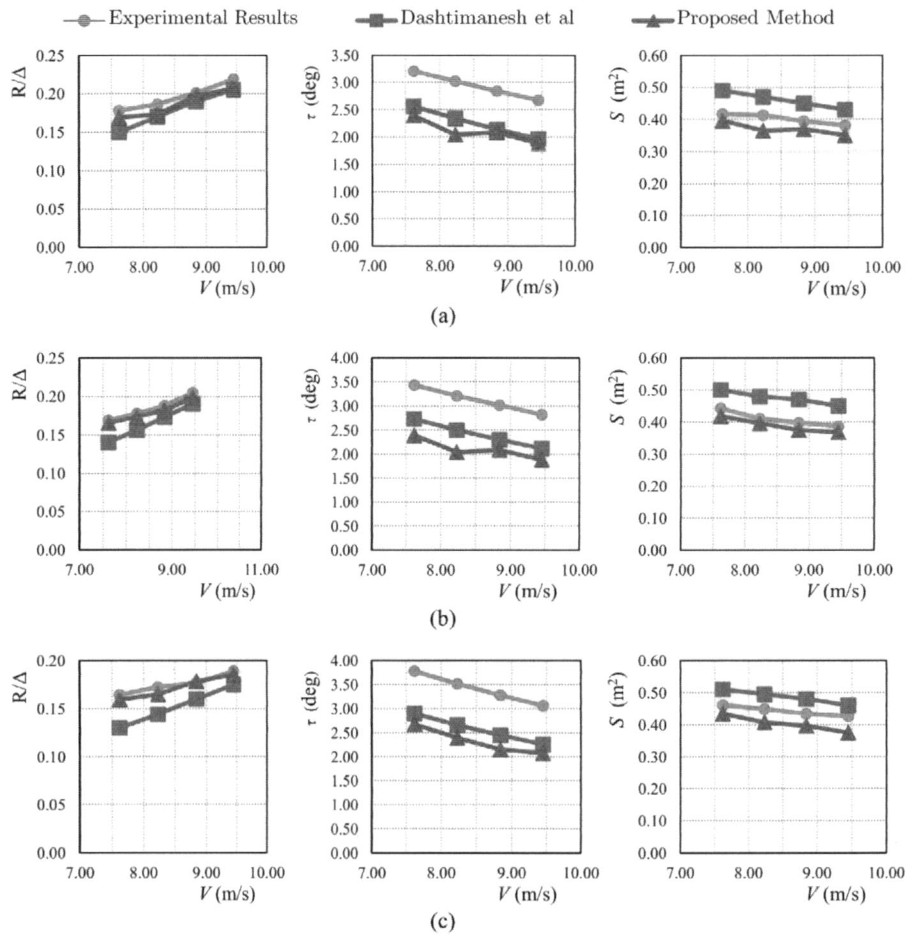

Figure 2: Comparison of result of presented method against experimental and mathematical results for case 7: (a) D = 38.55 kg, (b) D= 43.09 kg and (c) D= 47.63 kg [14].

The mean errors in prediction of trim angle, wetted surface and resistance were, respectively, 13%, 16% and 8%. To further evaluate the accuracy of developed 2D+T model for stepped hulls, we needed to know how much precise the mathematical model was for simulation of wetted surface pattern. For that purpose, Niazmand Bilandi et al. [15] depicted the wetted surface obtained by mathematical description of fluid flow and compared it with experimental data (Figure 3) for a stepped hull with swept steps while our mathematical model was designed for stepped hulls with transverse steps. Surprisingly, the mathematical model could predict the wetted surface relatively accurate. However, it was evident that side wetting could not be considered in our model as one of the possible wetting scenarios introduced in the Figure 1. We needed experimental data for various step shapes and configurations to be able to define an empirical formula for estimation of wetted surface.

It is also worth mentioning that another reason behind development of 2D+T method for two-stepped hulls was that this method could be further developed for performance prediction in waves. Accordingly, Niazmand Bilandi et al. [16] developed the 2D+T method for prediction of heave and pitch motions of two-stepped hulls in regular waves to make it clear how steps can influence the response of a planing vessel operating in waves. In fact, compared to a stepless design, the performance of a double-stepped body in waves can be different. It will be discussed in more details in the next section to compare 2D+T results against CFD simulations for performance prediction of stepped hulls in regular waves.

In the meantime, Niazmand Bilandi et al. [16] had mentioned that their developed mathematical model could be further developed to consider more degrees of freedom. Specifically, the roll motion could be considered in their dynamic model, which is likely to be very important for the case of high-speed vessels as they might show unstable behaviour in the transverse direction.

For that purpose, Niazmand Bilandi et al. [17] tried to model a heeled two stepped hull to study the effects of asymmetric conditions on wetted surface and in general, on boat performance. In lack of any experimental data, they compared results against CFD simulations. The quantitative comparison between the results of both methods showed that they predicted almost similar heeling moment, resistance and trim angle for a doubled-stepped planing hull. The comparison of wetted surface predicted by 2D+T method and CFD has been shown in Figure 4, for example. Again, it is seen that mathematical model cannot capture the side wetting, therefore, the formulation increases the area of wetting surface around the centerline to generate the required pitch moment. Consequently, it is again clear that we need a more reliable definition for wetting surface if we want to include roll motion in our formulation. Anyway, results of non-stepped and doubled stepped planing hulls were also compared against each other. Both 2D+T method and CFD demonstrated that the heeling moment of a double-stepped planing hull was smaller than a heeling moment of a non-stepped planing hull at early planing speeds, but, by the increase in speed, the heeling moment of doubled-stepped planing hulls became slightly larger.

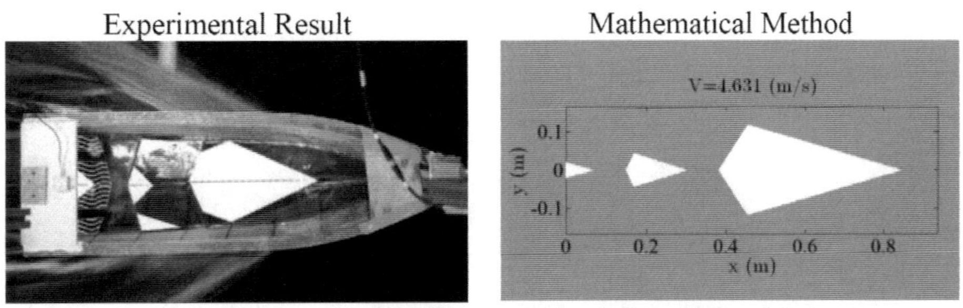

Figure 3: Comparison of the wetted area pattern found in experiments against what was observed in computations at beam Froude Number of 2.55 [15].

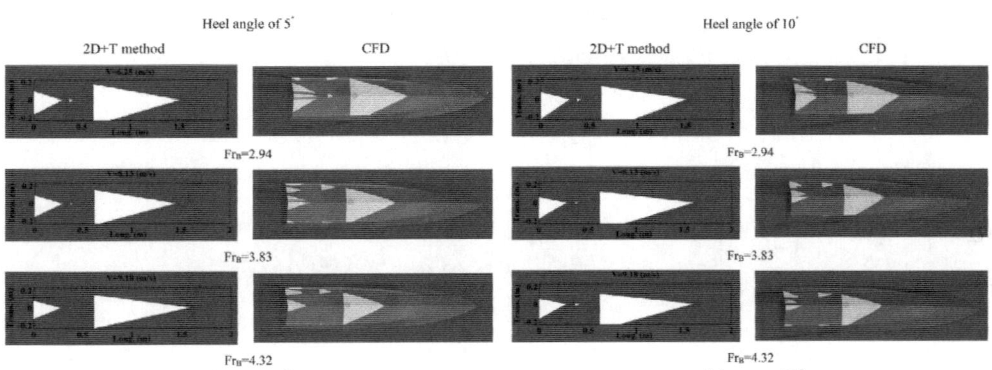

(a) $\varphi = 5^{\circ}$ (b) $\varphi = 10^{\circ}$

Figure 4: Comparison of predicted wetted surface of a heeled doubled-stepped planing hull by 2D+T method and CFD model at two heel angles and various Froude numbers [17].

4. CFD SIMULATIONS

As it was observed in previous section, we have developed different mathematical models for simulating the stepped hulls performance in calm water and waves as well as in manoeuvring motions. In parallel to those development, CFD models also developed by implementing CFD software Siemens PLM Star-CCM+ [18]. In one of our earliest CFD efforts, Dashtimanesh et al. [9] modelled a two-stepped hull with transverse steps in calm water by implementing the morphing mesh approach. Comparison with experimental data showed that the numerical solution was relatively accurate in terms of resistance, trim and sinkage prediction. Moreover, they investigated various hull characteristics, such as wetted length of keel, chine wetted length, spray angle and ventilation length, in their analysis. They suggested detailed simulation of fluid flow phenomena over the stepped hull as future studies.

For that purpose, several other CFD studies were also conducted by our team for considering stepped hulls performance in calm water such as Dashtimanesh et al. [19] and Dashtimanesh et al. [20]. One of our latest developed is a novel CFD model developed by Niazmand Bilandi et al. [21] in which the Unsteady Reynolds Averaged Navier-Stokes (URANS) equations are solved by employing Overset and Adaptive Mesh

Refinement (AMR) techniques in the framework of DFBI (Dynamic Fluid Body Interaction) approach. A detailed description of the computational domain and boundary conditions of the numerical tank is presented in Figure 5 where L is the length of the hull. They validated their CFD model (Figure 6) against extensive experimental data published by Vitiello et al. [22]. They tried to capture the details of fluid flow phenomena underneath the hull and provided valuable insights into the hydrodynamic performance of swept stepped planing hulls. However, the investigation was limited to a systematic series of stepped hulls with specific design parameters, and the use of the RANS equations has always its limitations in predicting unsteady flow phenomena. Authors concluded that accurate modelling of the wetted surface and ventilation length of stepped planing hulls is complex and requires careful consideration of the threshold value used for the fluid volume fraction and the iso-surface value used for free surface modelling, which can influence the accuracy of the results.

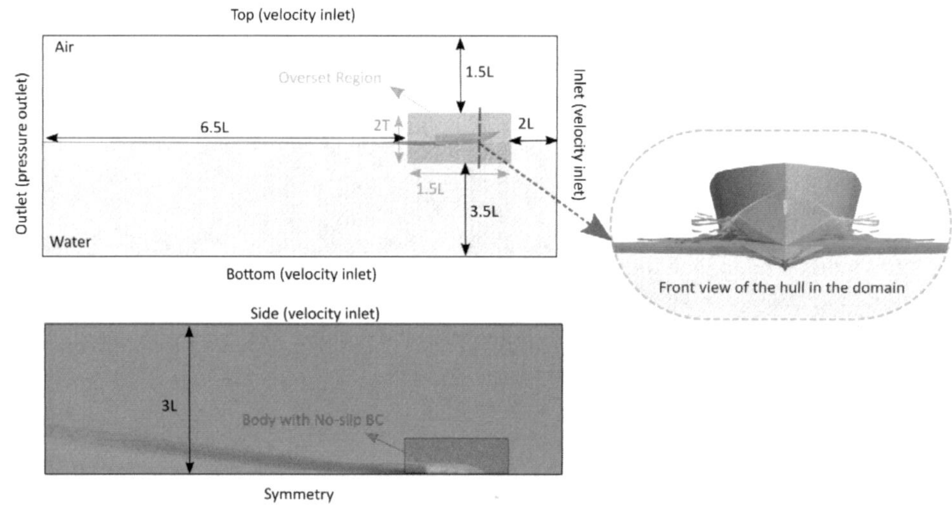

Figure 5: Computational domain and boundary conditions applied [21].

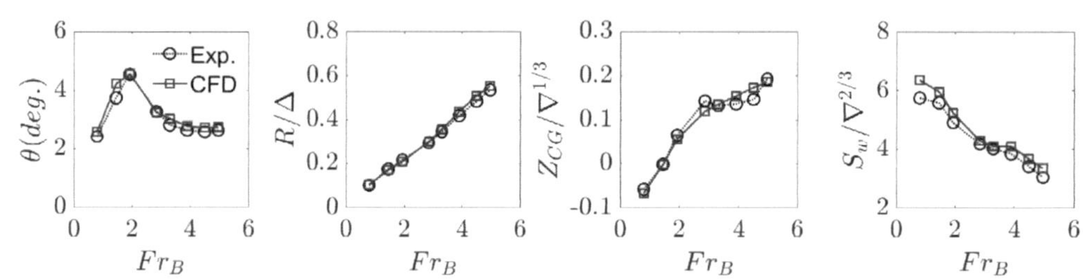

Figure 6: Comparison of the CFD results against the experimental results of Vitiello et al. [22], from the left to the right, are: dynamic trim, resistance, sinkage, and wetted surface, for the hull C08 [21].

Stepped hulls are operating in very high-speed regimes and consequently, are prone to dangerous turning motion. In fact, during the turning motion, when a stepped hull heels to port or starboard side, water fills the step and there is no ventilation on one side of the hull. In this situation, the distribution of hydrodynamic pressure and corresponding lift force over the related planning surface will be reduced significantly and consequently, the boat is prone to excessive transverse motions and even capsizing while throwing the crews/passengers out of the boat and/or leading to serious injuries. Therefore, it is very important to have a good understanding of what happens during the turning motion of stepped hulls from fluid mechanics point of view.

To address that challenge, capability of CFD solvers in investigating the heeled double stepped planning hulls was considered by Niazmand Bilandi et al. [17]. It could be considered as an initial step towards studying the manoeuvring motions of stepped hulls as one of the challenging aspects of stepped hulls design. The purpose of numerical experiments was to compare our 2D+T model to CFD simulations. It was observed that CFD and 2D+T method could provide similar results. Figure 4 showed the wetted surface of a heeled doubled-stepped planing hull predicted by 2D+T method and CFD model at two heel angles and various Froude numbers. As mentioned before, the results of this study proved that our fast 2D+T model can be further developed for modelling manoeuvring motions such as turning circle test. At the moment, our team are trying to model the turning motion of two-stepped hulls by developing/implementing a CFD model.

In addition to challenges in understanding and modelling of the manoeuvring motions of stepped hulls, their behaviour in waves has had its own specific challenges. It has been always believed or declared that stepped hulls have a very good seakeeping ability even in waves. However, there has been no scientific study on that. We have also tried to address this topic through our studies using CFD.

One of the very first CFD studies on stepped hulls in waves was conducted by Esfandiari et al. [23]. The problem was solved using the finite volume method in conjunction with volume of fluid method using CFD software Siemens PLM Star-CCM+ [18]. Two different wavelengths were considered in the computations. One wavelength was shorter than the boat length and the other one was longer than the length of the boat. The heave and pitch motions as well as vertical acceleration were investigated in comparisons. It was observed that at wavelength larger than the boat length, two-stepped hull could reduce the heave amplitude by 10–40%. In the meanwhile, pitch motion of stepped hull decreased by 18–32%. More importantly, two-stepped hull could reduce the vertical acceleration for the wavelength larger than the boat length. Authors concluded that if two steps with adequate height and location with respect to the transom are added to the bottom of high-speed planing hulls, their ability to operate in head sea improves. They mentioned that ventilated air at the bottom of vessels is still a source of concern regarding their stability in a transverse plane. It is worth noting that CFD computations in that study could not be validated because of lack of experimental results.

More recently, Niazmand Bilandi et al. [16] tried to develop their calm water CFD models to regular wave conditions. As mentioned before, in the lack of any experimental data, they compared their CFD results against their developed 2D+T method. It was observed that both CFD and 2D+T provide the same trends in most cases and results were completely comparable as it is seen in Figure 7. Niazmand Bilandi et al. [16] demonstrated that a double stepped design can decrease the heave and pitch responses of the vessel in the resonance zone (Figure 8). Vertical acceleration was also seen to be reduced when two steps were added to the bottom of the vessel. More importantly, mathematical data confirmed that wave-induced motions of a vessel are highly sensitive to the heights of steps. A vessel with a shorter front step has smaller vertical responses, especially over the resonance zone, while it may have larger vertical acceleration at high speeds. This is caused by the larger washed area of the middle surface, which can increase damping forces, while it allows larger wave forces to impact the vessel.

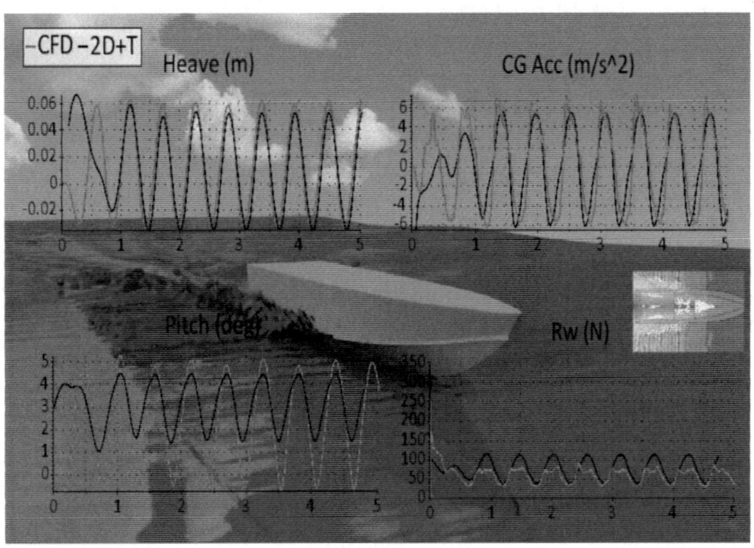

Figure 7: Comparison of CFD and 2D+T methods for simulation of motions of a two-stepped hull in regular waves.

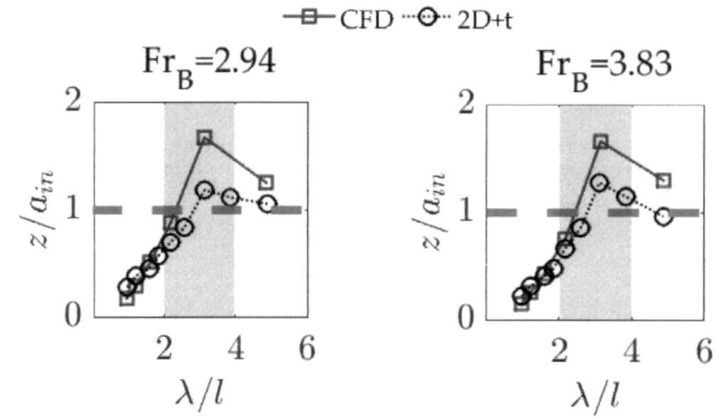

(a) Heave response,

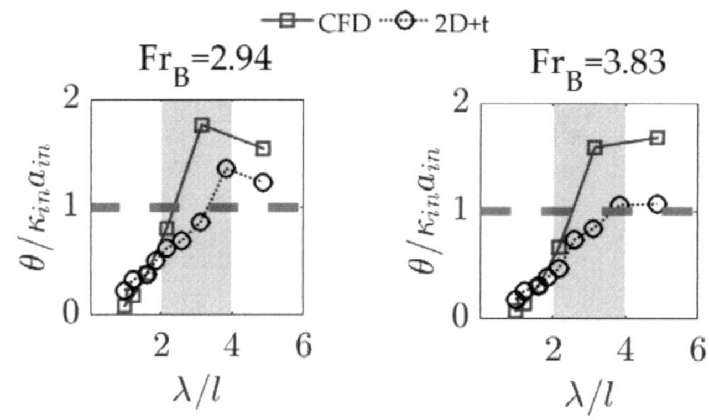

(b) pitch response,

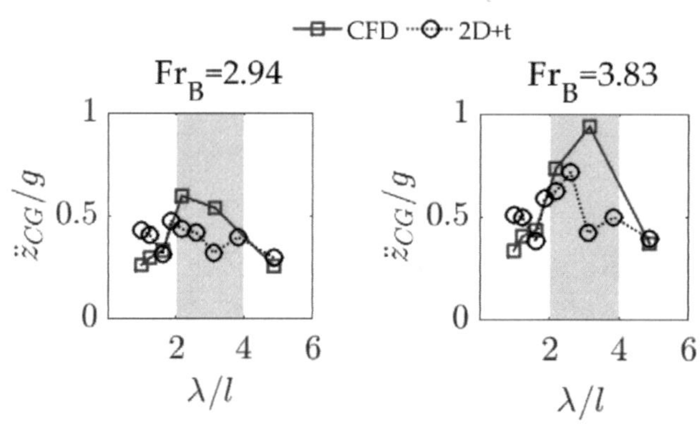

(c) vertical acceleration.

Figure 8: Heave and pitch responses and vertical acceleration of a two-stepped hull in waves [16]. The dashed red line shows the wave amplitude, reference line. Circle and square markers respectively show the 2D + t and CFD results.

5. EXPERIMENTAL TESTS

Although there have been many studies on calm water performance of stepped hulls, the effects of step height and position, the number of steps and step shape were still unknown. The focus of most of the previous studies has been on transverse steps while transverse steps are not the case suitable for industrial/commercial applications where forward or backward V-shaped steps, instead, are largely implemented. To tackle this challenge, Vitiello et al. [22] conducted a comprehensive towing tank measurements and presented a new systematic series of eight different stepped hull models with various step shapes, numbers, heights, and positions. The systematic series was developed based on a parent hull (RIB Mito 31) that was a traditional stepped hull with longitudinal spray rails and without an artificial bottom cavity or artificially inflated air in the cavity. The parent hull was a hard chine hull with two transverse steps with a forward V shape, with the

Center of Gravity (CG) located in the center of the surface between the first and second steps. All models had the same main geometric dimensions (keel line, chine line, deadrise angle, displacement, LCG, step shape, step angle) of RIB Mito 31, with a 1:10 scale ratio. The design parameters were defined based on a critical review of the literature related to the stepped hull design considering step number, step height, and longitudinal step position.

The towing tank tests were carried out at the naval basin of the Università degli Studi di Napoli "Federico II" Dipartimento di Ingegneria Industriale (DII) (Figure 9) in calm water at different speeds ($Fr_\nabla = 1.077\text{–}6.774$) and for three different static trim conditions. The calm water resistance, trim, sinkage, and wetted surface were measured. The towing tank test results were completely released in the paper that could be used for validation and verification studies. Moreover, a detailed uncertainty analysis was conducted by authors following the ITTC procedure [24] for uncertainty analysis in resistance towing tank tests (7.5-02-02-02). The measures with the highest uncertainty were the sinkage and wetted surface. The wetted surface uncertainty is mainly related to the spray areas which were difficult to be estimated based on the recorded photos and videos.

All models had been built with a transparent bottom to visualize the complicated fluid flows underneath the stepped hulls. It was observed that some vortices developed into the aft body region behind the step and partially continued downstream in the wake. For more details, authors reported the fluid flow and wetted surface pictures of all stepped hull models in trimmed forward condition at whole speeds. Wetted surfaces of one of the stepped hull models at the different tested speeds has been shown in Figure 10. Analysing the wetted surface of various models at different speeds will provide a better understanding of what happens underneath the stepped hulls and accordingly, it will be possible to improve the accuracy of our developed mathematical models mentioned in previous sections. In fact, accuracy of mathematical models based on 2D+T method highly depend on accuracy of wetted surface predictions.

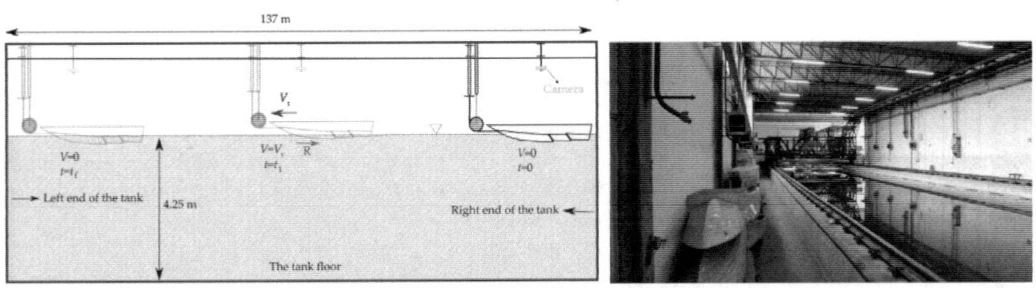

Figure 9: The schematic of conducted resistance tests in the University of Napoli "Federico II" towing tank.

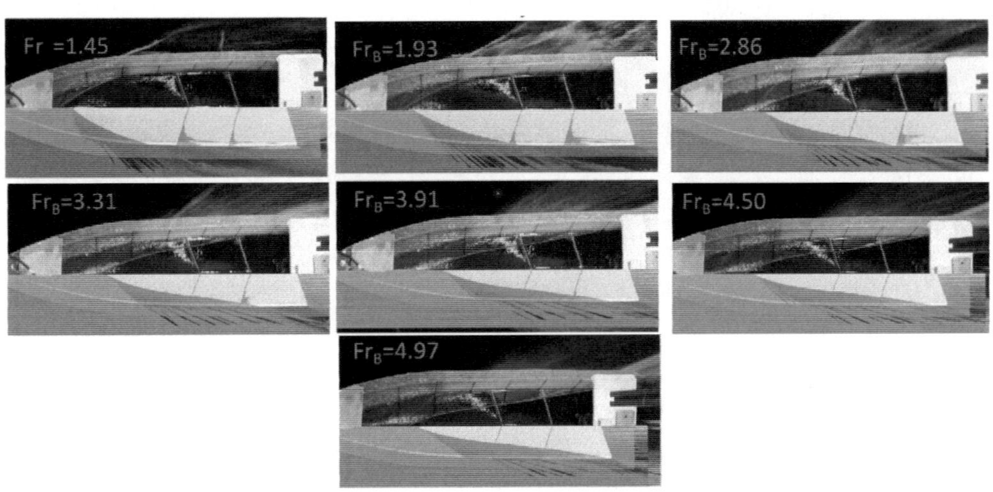

Figure 10: Wetted surfaces of one of the stepped hull models at the different tested speeds [22].

6. CONCLUSIONS

In this paper, it was tried to address our recent efforts towards sustainable fast boats by focusing on stepped hulls as one of most promising options in the market. Stepped hull design has always been a complicated task for boat designers and there have been many challenges and research questions related to their design. These complexities are for example including the effects of step position, step height, step shape and number of steps on boat performance in calm water, in waves and during manoeuvring motions.

Accordingly, it was described in the paper how we have tackled with these challenges through our previous mathematical, CFD and experimental studies. Our developed mathematical based on 2D+T method can model stepped hull with various step characteristics in calm water and in regular wave. We have validated 2D+T method for stepped hull by comparison the obtained results against both CFD simulations and towing tank experiments. We are now working on developing our method for inclusion of irregular waves in the 2D+T model. Our mathematical tools can also model the stepped hulls in heeled conditions, but not yet extended to roll motion prediction. These mathematical models and developed computer programs can be considered as a design tool for the early-stage design.

Moreover, it was explained how we have developed various CFD setups for modelling stepped hulls in calm water and waves. We are now trying to develop CFD models for performance prediction in irregular waves and in manoeuvring motion. However, even by implementing high performance computing clusters, the computational time is too high. In the meanwhile, CFD simulation could reveal many physical details underneath the stepped planning hulls that has led to better understanding of their design. This acquired knowledge will be implemented in our mathematical models to improve their accuracy.

In addition to our mathematical and CFD efforts, we have developed a comprehensive systematic series for stepped hulls by considering various step shapes, step numbers, step height, and longitudinal position of steps. During the experiments, we have captured the physical phenomena underneath the hull thanks to our developed transparent hulls. The experiments on this systematic series will be extended by testing it in regular waves. We expect that these efforts will reveal hidden aspects of stepped hulls design. It is also worth mentioning that we aim to develop a machine learning model for designing stepped hulls. An initial model has already been developed for the simple case of Fridsma hull. While nowadays, most people are again interested in hydrofoil boats, we believe that stepped hulls will survive because of their wide applications in the leisure boat industry as the main stimulator of boating market.

7. ACKNOWLEDGEMENTS

Author would like to express his sincere appreciation and thankfulness to his colleagues, Sasan Tavakoli, Rasul Niazmand Bilandi, Simone Mancini, and Luigi Vitiello, because of their efforts and contributions in our previous research addressed in this paper.

8. REFERENCES

1. HELCOM, 'State of the Baltic Sea – Second HELCOM holistic assessment 2011-2016', *Baltic Sea Environment Proceedings 155*, 2018.

2. IMO, 'Initial Imo Strategy on Reduction of GHG Emissions from Ships', *Annex. 11, Resolution MEPC.304(72)*, 2018.

3. MOLDANOVÁ, J., FRIDELL, E., MATTHIAS, V., HASSELLÖV, I-M., JALKANEN, J-P., TRÖLTZSCH, J., QUANTE, M., JOHANSSON, L., KARL, M., MALUTENKO, I., YTREBERG, E., ERIKSSON, M., SIGRAY, P., KARASALO, I., PELTONEN, H., HASENHEIT, M., GRANHAG, L., MAWDSLEY, I., AULINGER, A., PIOTROWICZ, J., 'Sustainable Shipping and Environment of the Baltic Sea region: Final Report. SHEBA', 2018.

4. HIRDARIS, SE., CHENG, F., 'The role of technology in green ship design', *11th International marine design conference*, 2012.

5. DASHTIMANESH, A., HIRDARIS, S., GHAEMI, M.H., 'Green Cruising for High-speed Small Craft in The Baltic Sea', 2021. (https://interreg-baltic.eu/project/green-small-craft/)

6. DASHTIMANESH, A., GHAEMI, M.H., WANG, Y., KARCZEWSKI, A., NIAZMAND BILANDI, R., HIRDARIS, A., 'Digitalization of High Speed Craft Design and Operation Challenges and Opportunities', *Procedia Computer Science,* 2022.

7. DE LUCA, F., PENSA, C., 'The Naples warped hard chine hulls systematic series', *Ocean Engineering*, 2017.

8. DI CATERINO, F., NIAZMAND BILANDI, R., MANCINI, S., DASHTIMANESH, A., DE CARLINI, M., 'A numerical way for a stepped planing hull design and optimization', *Technology and Science for the Ships of the Future*, 2018.

9. DASHTIMANESH, A., ESFANDIARI, A., MANCINI, S., 'Performance prediction of two-stepped planing hulls using morphing mesh approach', Journal of Ship Production and Design, 34(3), 236-248, 2018.

10. DASHTIMANESH, A., TAVAKOLI, S., SAHOO, P., 'A simplified method to calculate trim and resistance of a two-stepped planing hull', *Ships Offshore Struct*, 12, S317 – S329, 2017.

11. SVAHN, D., 'Performance prediction of hulls with transverse steps', *Stockholm: Marina System Centre for Naval Architecture, KTH University*, 2009.

12. DANIELSSON, J., STRŲMQUIST, J., 'Conceptual design of a high-speed superyacht tender hull form analysis and structural optimization', Master Thesis, *Stockholm: Marina System Centre for Naval Architecture, KTH University*, 2012.

13. SAVITSKY, D., Morabito, M., 'Surface wave contours associated with the forebody wake of stepped planing hulls', *Marine Technology*, 47 (1), 1 – 16, 2010.

14. NIAZMAND BILANDI, R., DASHTIMANESH, A., TAVAKOLI, S., 'Development of a 2D+T theory for performance prediction of double-stepped planing hulls in calm water', Proceedings of the Institution of Mechanical Engineers, *Part M: Journal of Engineering for the Maritime Environment*, 233(3), 886-904, 2019.

15. NIAZMAND BILANDI, R., VITIELLO, L., MANCINI, S., NAPPO, V., ROSHAN, F., TAVAKOLI, S., DASHTIMANESH, A., 'Calm-water performance of a boat with two swept steps at high-speeds: Laboratory measurements and mathematical modeling', *Procedia Manufacturing*, 2020.

16. NIAZMAND BILANDI, R., TAVAKOLI, S., DASHTIMANESH, A., 'Seakeeping of double-stepped planing hulls', *Ocean Engineering*, 236, 109475, 2021.

17. NIAZMAND BILANDI, R., DASHTIMANESH, A., TAVAKOLI, S., 'Hydrodynamic Study of Heeled Double-Stepped Planing Hulls Using CFD and 2D+ T Method', *Ocean Engineering*, 196, 106813, 2020.

18. Siemens PLM, 'STAR CCM+ User's Guide Version 17.02.007', 2022.

19. DASHTIMANESH, A., ROSHAN, F., TAVAKOLI, S., KOHANSAL, A., BARMALA, B., 'Effects of step configuration on hydrodynamic performance of one-and doubled-stepped planing flat plates: a numerical simulation', *Proceedings of the Institution of Mechanical Engineers, Part M: Journal of Engineering for the Maritime Environment*, 2019.

20. DASHTIMANESH, A., TAVAKOLI, S., KOHANSAL, A., KHOSRAVANI, R., GHASSEMZADEH, A., 'Numerical study on a heeled one-stepped boat moving forward in planing regime', *Applied Ocean Research*, 96, 102057, 2020.

21. NIAZMAND BILANDI, R., DASHTIMANESH, A., MANCINI, S., VITIELLO, L., 'Comparative study of experimental and CFD results for stepped planing hulls', *Ocean Engineering*, 2023.

22. VITIELLO, L., MANCINI, S., NIAZMAND BILANDI, R., DASHTIMANESH, A, DE LUCA, F., NAPPO, V.,. A Comprehensive Stepped Planing Hull Systematic Series: Part 1 - Resistance Test. Ocean Engineering, 266, p.112242, 2022.

23. ESFANDIARI, A., TAVAKOLI, S. AND DASHTIMANESH, A., 'Comparison between the dynamic behavior of the non-stepped and double-stepped planing hulls in rough water: a numerical study', *Journal of Ship Production and Design*, 36(01), 52-66, 2020.

24. ITTC, 2002b. Recommended Procedures and Guidelines, 7.5-02-02-02.

9. AUTHORS BIOGRAPHY

Abbas Dashtimanesh is an Assistant professor of Naval Architecture at KTH Royal Institute of Technology. He is working in department of Engineering Mechanics focusing on ship motion prediction methods, hydrodynamics of high speed craft, application of machine learning in ship hydrodynamics, and sustainable shipping through digital twins opportunity.

GREEN FUELS OPTIONS STUDY FOR HIGH SPEED MARINE VESSELS

C Saunders, J Huber and **J Lahiri**, DSTL, UK

SUMMARY

The MOD operates a range of small boat based capabilities across a mix of roles. The replacement of these capabilities will, in particular, present unique design challenges when attempting to meet net-zero targets while maintaining their desired military capability.

This paper reviews the approach and outputs of a study looking at green fuels options for these platforms. It investigated the different options that could be utilised to propel future small naval platforms while aiming to achieve net zero emissions by 2050. It considered options including ammonia, hydrogen, battery based hybrid systems, and the use of synthetic hydrocarbons.

These options are considered against two key design conditions, a logistically constrained platform at 10 tonnes and a purely capability based view with minimal 'iron triangle' (payload vs. speed vs. range) constraints.

Based on our findings the study draws conclusions and makes recommendations for further research.

NOMENCLATURE

A_p	Chine Area (m^2)
$Ap/\Delta^{2/3}$	Hull Loading coefficient
BOA	Beam Overall (m)
B_{px}	Chine Beam (m)
E_T	Transport efficiency
g	Gravity ($m\ s^{-2}$)
LOA	Length Overall (m)
L_p	Chine Length (m)
P	Power (kW)
ULF	Useful load fraction
V	Velocity ($m\ s^{-1}$)
Δ	Displacement (m^3)
Δ_{FL}	Displacement Full load (m^3)

1. INTRODUCTION

Over the last 50 years, there has been an increase in awareness that the pollution humanity is putting into the environment is negatively altering our world, International, national and local emission regulations have been enacted to limit the release of harmful (both to health and the environment) emissions. These are increasingly supported by political, financial and societal pressure to reduce harmful emissions. Reducing emissions is required across all sectors of industry and society, with one of the key strategies to reduce emissions through the mass electrification of the ground transportation sector.

Research has focused on this approach and the enabling technology has been developed to a level where the electrification of the consumer car market is now feasible. With the car market leading the way, wider industry is looking to exploit and expand the technology to other sectors of the transportation market such as light aircraft, boats and small craft.

the conclusions within this paper do not represent the formal views of the UK Government or Defence, rather reflects the outputs of an underpinning study completed to help decision makers to make informed decisions.

1.2 PURPOSE

The primary question this study aims to answer is:

"What options exist to support MOD's 2050 net zero emission target without sacrificing capability for small naval craft on tasking?"

Whilst exemplars will be provided focusing in on small craft at 10 tonnes, the analysis scales easily to be applicable across all vessel sizes[1]*. Thus findings could be applicable across the whole future fleet depending on vessel duty cycle*[2]*.*

The study was conducted under the Specialist Systems programme and was funded by UK's defence Chief Scientific Advisor (CSA) Research funding.

Within this study green means low greenhouse warming potential when compared to the other options.

1.3 METHOD

The method chosen for this study is to calculate the total energy required for both the production of each fuel type considered and then used within the fuel burnt on mission. This approach was chosen above focusing solely on the energy content for each fuel, as this creates a fairer overall comparison of net CO_2 emissions on a 'well-to-wake' (or equivalent) basis. This study assumes that all fuel generation and synthesis is achieved with 100% green energy sources, hence negating the potential impacts of the use of different mixes of energy sources (Nuclear, Coal, Gas, Wind etc.).

The study determines which fuel option is *"greenest" by looking at the lowest energy requirement to generate/manufacture the fuel required to meet the mission.* The one exception to this is synthetic diesel operating without a scrubber. This is because there are particulate and Nitrous Oxides (NOx) emissions that need to be accounted for in the overall emissions picture.

This report does not consider cradle to grave emissions of a small boat as defined in "Defence Carbon Baseline Methodology and Calculation" (1). The only emissions considered in this study are those related to the fuel required for the operation of the vessel (excluding hotel electrical loads).

1.3.1 Assumptions

The analysis in this report is underpinned by the following key assumptions.

- All fuels are made from electrical energy (assumed to be produced by a 100% renewable source).
- Vessel design is limited to hard chine mono-hulls due to information available, MOD logistics constraints and the current design direction of most MOD small vessels.
  - No consideration has been given to the impacts of fuel options on vessel interactions with other systems, infrastructure and logistics facilities.
  - Designs are based on best assumptions for likely performance versus system weight.
- No consideration has been given to relative costs, access to or location of feed-stocks or fuel production facilities, or to logistics issues (e.g. time to re-fuel).

---

[1] As vessel size increases and operational Froude number decreases, the error bands of the analysis will start to expand. This includes both planning and displacement craft.

[2] The main differences in the analysis between different craft sizes are an expanded payload definitions, added mass due to long term habitability requirements and hotel electric loads (become a significant power draw compared to main propulsion power).

- It is assumed all considered technology can be marinised, without impacting their weight.
- The use of synthetic hydrocarbons - this simplifies analysis and assumes the removal of sulphur dioxide (SOx) (i.e. with no significant sulphur content). This will result in increased fuel costs, and does not remove emission of NOx.
- Wind and Solar as a primary or secondary on board power solution are both considered incompatible with the requirements of the user community for this report and as such are not considered feasible options. This is due to the uncertainty of solar and wind power availability at point of use.
- Direct drive between the power unit and propulsors, where possible due to efficiency of the transmission and reduction in power losses. Gearbox options have the same efficiency and weight, whether a planetary, continuously variable transmission (sometimes referred to as CVT) or meshed option is selected.
- Vessel control systems are designed to maximise the efficiency of each powertrain option and therefore the net impact on performance of these systems is negligible. This limits the number of different duty cycles to be considered.
- IMO tier 3 (which currently includes the North Sea, Baltic Sea, North America and Caribbean Sea) implementation timeframe is currently unknown and will likely go global at some point. This will make it increasingly difficult for current systems to meet emissions standards as they are rolled out more widely.
- Steady state speeds are assumed, so no consideration is given for additional loads due to acceleration or hotel loads. This assumption provides the lowest bar of a criteria to determine whether a particular concept is feasible. If this threshold can be met, further investigation is required in order to balance the requirements with the specific details of the application.
- Safety and Survivability risks to the platform from carrying the fuels suggested is outside of scope.
- All technology can operate reliably in the small boat environment. Depending on results, some technologies may need to be tested and investigating further.
- This study only considers lowest energy requirement for a vessel on tasking. This does not include cradle to grave emissions.

2. ALTERNATIVE ENERGY SOLUTIONS

The following sections summarise the main energy comparison for alternative fuel sources. This should help MOD and the Royal Navy (RN) explore the reality of achieving net zero in a small high speed vessel.

All of these solutions have potential to have no greenhouse gas emissions, if 100% "green" energy sources are used to power fuel synthesis, transportation, storage and delivery. An example would be, if the correct production process is used, CO_2 is taken from the atmosphere to generate synthetic diesel (using green electricity), so when the synthetic diesel fuel is combusted, overall CO_2 generation is neutral. Therefore another metric is needed to compare the desirability of alternative energy solutions, which is developed in subsequent chapters. It is important to note, however, that currently no energy source can be considered to be 100% green. This is because all sources of energy in the current national grid that could be used to synthesise fuels have emissions higher than are consumed on a day to day operational level[3].

2.1. ALTERNATIVE ENERGY INPUT COMPARISON

Whilst it is challenging for a fuel to achieve net zero in operation, the challenge of achieving net zero from point of generation is arguably even greater. All the fuel or energy storage sources considered in this study

[3] Hopefully this will change however trying to resolve this issue is beyond the scope of this study.

require considerable manufacturing, processing and input energy before they are even deployed to a small boat. The following section addresses the energy efficiency of the generation process for each of the options in this study to enable comparisons to be made on a per mission basis.

2.1.1. Energy requirement to generate fuel

This section looks at the energy usage per tasking for each of the "green" fuel options considered in this study. Whilst it considers the energy input to generate the fuel as part of the process it does not consider the energy required to manufacture the energy storage system. As a result the fuel cell and synthetic E-fuel options both include the energy usage to manufacture the fuel in their efficiency calculation whilst the battery electric option only looks at the losses from the point of electrical generation, thus considering transmission losses between the generation sources and charging infrastructure.

Figure 1 shows the proportion of input energy that is actually usable on the platform for each of battery electric vehicles, fuel cell vehicles and a diesel engine running on synthetic diesel.

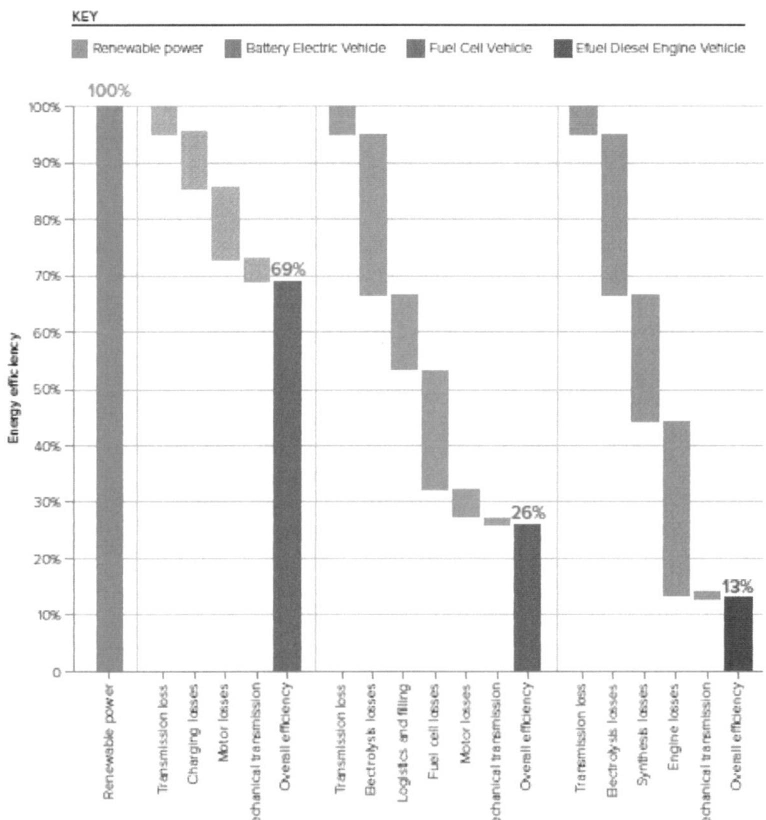

Figure 1: Waterfall chart of pathway efficiencies for low-carbon automotive transport (2)[4]

Figure 1 is an example from the automotive industry and as such the plot would look slightly different for a small boat, with all of the systems becoming more energy efficient due to the reduction in mechanical transmission losses in the marine environment. However, it does not change the order of the graph. Battery electric vehicles still require the smallest energy input followed by fuel cells and then Synthetic diesel.

[4] Synthetic diesel graph is assuming carbon capture. The uncertainty as to how the CO_2 is captured and from what source which could significantly change the percentage. Graph is unlikely to be straight from atmosphere.

3. TECHNICAL SOLUTION COMPARISON

Inputs can be found in Table 1.

3.1 METHODOLOGY

The following sets out the analysis method applied to assess the performance of different low energy/emissions fuelling options in small boat applications. In each section the methodology applied is the same however the assessment metrics to determine the relative performance of the vessels changes. In section 3.2 the key assessment metric is the initial energy input required to generate enough fuel to complete the mission.

This acts as a measure of how truly "green" the fuel may be by assessing it throughout its lifespan from generation rather than just considering the end use operational emissions coming out the boats exhaust. Section 3.3 then assesses the impact on capability that the introduction of "green" fuels is likely to have on the small boat fleet. This assumes that the boats operate under the same logistics constraints as current and future planned assets and compares the performance of the alternative fuels to diesel engines designed to meet current IMO and EPA (Environmental Protection Agency) emissions regulations that could be run on either natural or synthetic diesel.

3.2 INITIAL ENERGY INPUT

To consider the problem in the round instead of just looking into the lightest and most capable powering solutions, this section aims to consider the approximate amount of energy required to complete a tasking relating to a small boat. The energy required to create and dispose of the vessel or transport and supporting logistics is not considered.

Assumptions are that speed and payload are fixed. Vessel displacement and range are allowed to vary as required to ensure that a vessel design is viable.

3.2.1 Analyses based on current technology

This analysis is formulated using the Useful Load Fraction (ULF) and Transport Factor (TF) equations.

Useful Load Fraction is defined below. Using the desired range as a requirement, the amount of payload for a given speed, range and fuel consumption rate can be calculated.

$$ULF = \frac{Payload + Fuel}{\Delta_{FL}}$$

TF is defined by the following equation:

$$T_F = \frac{\Delta * g * V}{P}$$

Table 1 shows the input parameters used for the model when looking at the energy input required to generate the fuel.

Table 1. Input into energy required to generate fuel.

	'Burn Rate' of Energy	Storage + Fuel Weight Assumption multiplier	Power Plant Power Density	Driving Force Output/ "Green" Energy Input (2)
Battery	2.50 kg/kWh[5]	1	5.00 kW/kg	72.7%
Hydrogen Fuel cell	0.08 kg/kWh	Compressed Hydrogen gas storage - 13.33 Transcritical Hydrogen storage - 5[6]	1.61 kW/kg	Compressed Hydrogen Gas Storage - 27.3% Transcritical Hydrogen Storage (Assumed) - 13.7%
Diesel engine	0.22 kg/kWh	1.06	0.50 kW/kg	13.7%

3.2.2 Discussion

Sprint speed has the biggest single effect on total energy requirements. A need to maintain current achievable speeds (i.e. based on current technology) would favour the use of hydrogen or retention of diesel technology using synthetic fuels.

The use of battery based technology to power a small boat has significant weight impacts. These impacts are mostly driven by sprint speed, however range also has an impact. If payload is increased, the use of battery technology will be "greener" compared to synthetic diesel for longer ranges.

Battery technology could be the most suitable solution for a ship's sea boat due to its ability to support the host platforms to be "green"[7].

The use of compressed hydrogen is "greener" than synthetic diesel up to an approximate range of 250 nautical miles. Increasing payload does cause compressed hydrogen technology to be "greener" compared to synthetic diesel for longer.

The energy requirement for the generation, pressurisation and cooling of hydrogen into a transcritical state is comparable to synthetic diesel. Therefore this option can't really be defined as "green" until speeds are greater than 40 knots, requiring short operational ranges, or the vessel's operational emissions are considered in isolation.

3.2.3 Sensitivity of results to technology improvements

Many reports state that a 20-30% fuel burn saving (grams per kilowatt of power per hour) is possible compared to current diesel engines (4). If these improvements were delivered, then this would significantly drive the most "green" solution towards being synthetic diesel for military vessels. This efficiency improvement comes at the cost of engines requiring pre-heating[8].

Hydrogen storage weight is the biggest issue that holds back hydrogen from requiring less energy than synthetic diesel. However if compressed hydrogen storage could reach 20% hydrogen by weight when full, then this would be the "greenest" option for most small boats, so long as overall displacement is not a

---

[5] 400 W/kg battery density

[6] A storage weight of 5x times the weight of hydrogen is technical achievable but highly challenging. There is significant risk that regulators will not approved storage for this type within the maritime domain.

[7] This is due to ship being able to run smaller electrical power margins (the battery in sea boat could be used to manage the margin), also the standard tasks that the sea boats are expected to complete.

[8] Significant development investment would be required from either an OEM or MoD to deliver upon these potential fuel burn savings.

constraint. The exception to this, is specific niche (short duration/range) roles where battery technology would be "greener".

Significant improvements in battery power density need to occur for a vessel powered by batteries to be "green" over ranges that military vessel require. Currently a doubling of battery power density has a large impact on vessel displacement thereby significantly improving logistics and energy requirements to complete a tasking. If a 3-4 times increase in current power density is achieved, then it would be worth reconsidering batteries as a primary energy source, not just for use in a hybrid configuration.

Future efficiency improvements in fuel synthesis and processes for each of the options are being widely considered. In particular, research is ongoing to reduce the energy needed to crack water into hydrogen and oxygen. This would improve the efficiency of both hydrogen and synthetic diesel production.

3.3 LOGISTICAL CONSTRAINED ASSETS (10 TONNE CASE STUDY)

A case study of a 12 m Length overall (LOA) deep V mono-hull with a 10,000 kg displacement was conducted. In order to show the trade-offs in range and payload for four propulsion technologies: a IMO tier III diesel engine system, EPA tier IV diesel engine system driving a gearbox and propeller shaft, and, a battery powered, and a hydrogen fuel cell powered electrical drive chain.

In this analysis the displacement and speed will be considered fixed to some degree and payload varied. It does not consider the energy required for creation of the fuel, but what the "Iron Triangle"[9] trade off looked like for different fuels.

Vessel particulars are shown in Table 2.

Table 2. 12 m Vessel Case Study General Particulars

Length Overall - LOA (m)	12
Chine Length - L_p (m)	10.2
Beam Overall - BOA (m)	3.2
Chine Beam - B_{px} (m)	3.2
Chine Area – A_p (m^2)	27.1
Displacement - Δ (kg)	10,000
Bottom Loading - $Ap/\Delta^{2/3}$	5.9

For a boat, many of the trade-offs are related to weight. The useful load concept, ULF, demonstrates this concept by showing that for a given weight of vessel, there is a percentage of the total weight of the craft that can be allotted to payload and fuel for a given speed with the rest of the weight being reserved for the hull: structure, machinery, etc. How the trade-offs are conducted will determine how much range or payload one can achieve.

For mission sets where range is a higher priority, the amount of fuel required to achieve the range at the desired speed will be most important. For mission sets where the payload value is fixed, the weight remaining for fuel will dictate the ranges possible at the desired speeds. The payload needed for a mission set is generally dictated by the user community that is defining the boat's mission. Since this value is variable depending on the user community, and mission, the study used range as the controlling value for the mission and show how the speeds and the ranges affect the amount of weight available for payload.

Figure 2 shows how the payload changes for the four technologies at speeds of 30 and 40 knots for ranges of: 50, 100, 200, 300, 500, 1000 nm. The "green" solid and dashed lines are the payload baseline configuration: an International Maritime Organisation (IMO) Tier III compliant diesel engine at 30 and 40 knots, respectively.

---

[9] The iron triangle defines the design space (at a specified displacement) in which the vessel operates. It shows the trade-offs between speed, payload and range based on the overall system requirements at a specific design point. The iron triangle can be expanded by increasing vessel displacement.

Where another line intersects the "green" baseline solid or dashed line, is the location where the comparative technology has equal range as the IMO Tier III diesel engine. The payload and range values have been summarized in Table 6 and Table 7 to highlight maximum payload values for a given range and speed for the four technologies examined.

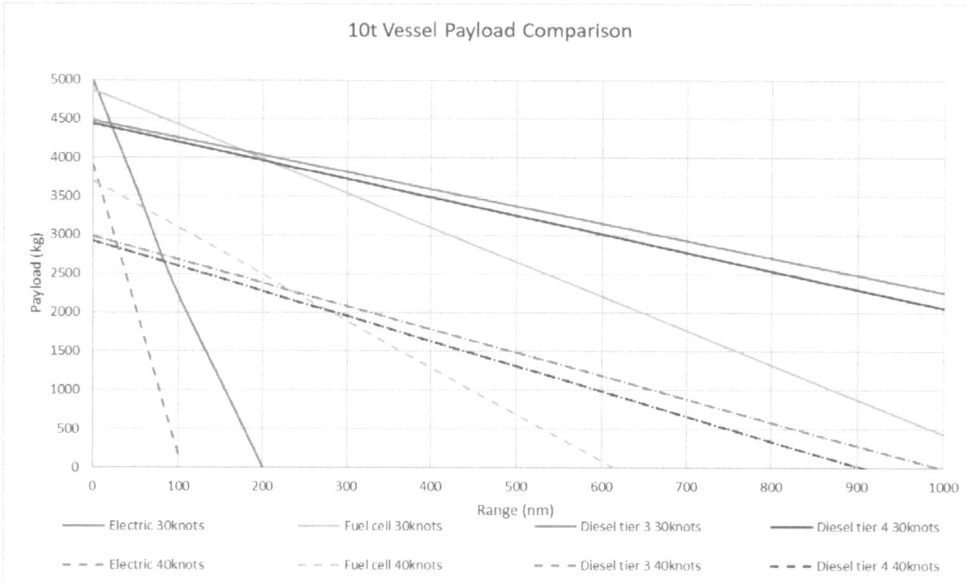

Figure 2: Payload Variations vs Range for Propulsion Systems at 30 and 40 knots

Figure 2 shows that there are "green" solutions, electric and hydrogen, that could be used for short ranges and provide the same payload capacity equivalent IMO Tier III diesel system. The ranges and payload capacity for an electric and hydrogen fuel cell compared to an IMO Tier III Diesel engine are summarized in Table 5.

Table 3. Range and Payload for "Green" Propulsion Systems with Equivalent Capability to IMO Tier III Diesel System

	30kts		40kts	
	Payload (kg)	Range(nm)	Payload (kg)	Range(nm)
Fuel cell 2kw/kg	4424	176.3	3090	235.1
Electric	2217	20.1	146	26.8

Table 4 and 5 show how the amount of payload changes with range at 30 and 40 knots. While hydrogen and electric options are viable, the amount of payload that can be carried is less than an IMO Tier III or EPA Tier IV diesel system. Where the systems show a dash for a given range, the amount of weight that is available for payload and fuel has been entirely used by the propulsion system and energy store/fuel and no weight is available for payload.

Table 4. Payload Variations for 6 Ranges at 30 knots

	Range (nm)						
	50	100	200	300	500	750	1000
Propulsion System	**Available Payload (kg) at 30 Knots**						
IMO tier 3 Diesel	4370	4260	4041	3821	3382	2833	2285
Hydrogen fuel cell 2kW/kg (transcritical storage)	4649	4429	3988	3548	2667	1566	465
Hydrogen fuel cell 2kW/kg (compressed gas storage)	4282	3695	2520	1346	-	-	-
Electric (battery 400w/kg)	3611	2235	-	-	-	-	-

Table 5. Payload Variations for 6 Ranges at 40 knots

	Range (nm)						
	50	100	200	300	500	750	1000
Propulsion System	**Available Payload (kg) at 40 Knots**						
IMO tier 3 Diesel	2854	2707	2412	2118	1529	793	56
Hydrogen fuel cell 2kW/kg (transcritical storage)	3402	3107	2516	1925	744	-	-
Hydrogen fuel cell 2kW/kg (compressed gas storage)	2910	2122	547	-	-	-	-
Electric (battery 400w/kg)	2064	218	-	-	-	-	-

3.3.1 12M, 10 tonne case study summary

There are "green" options that provide the same capability as an IMO Tier III Diesel engine in some circumstances; however, the ranges where these systems provide equivalent capability are limited and may not be of interest to all user communities. "Green" options exist for ranges of interest, but the payload values will be considerably less than current capability for the same size of craft.

Batteries are too heavy to make it worth considering an all-electric vessel. However, batteries can be useful in a hybrid solution[10].

3.4 DISCUSSION

Both of the analyses in 3.2 and 3.3 show the scale of the challenge associated with moving from current IMO Tier III compliant diesel engines over to net zero equivalents. The key challenge, as can be seen in 3.3, is the energy density of the net zero technical solutions and the corresponding increases in propulsion weight that this causes. This means that, due to the impact of the small boat iron triangle, the net zero equivalents will have a smaller mission endurance or payload than the vessels they are aimed at replacing.

Currently the single best option is to just travel slower in the short term. This would reduce the amount of power required to move a vessel over a set distance. However, it is recognised that reducing speed may not be possible for most military tasks.

Secondary driving factor that affects a high-speed boats emission is based on range that the vessel is aiming to cover.

4. CONCLUSIONS

4.1 LOWEST ENERGY TECHNOLOGY ON A PER TASKING BASES

The key driver for selecting the "greenest" fuel for high-speed vessels is maximum range.

- Batteries are the "greenest" solution to travel <150 nm based on current COTS technology. The range at which batteries are "greenest" increases with battery power density. This holds true so long as both size and weight limitations of a vessel are not driving requirements of the design, regardless for battery power density.

[10] This is depending on vessels duty cycle and therefore could reduce capability overall capability and or increase emissions.

- Hydrogen fuel cell powered craft would be possible however endurance (range), fuel risk, energy requirements and size of craft are areas significantly affected. There are other reasons this technology might be considered such as, for example, silent running.

    - Improvements in hydrogen storage under pressure needs to at least double fuel/ storage container capacity on a unit weight basis. In addition, both RN, RFA and RAF assets need to work on what level of risk can be carried with storage of hydrogen on-board.

- Synthetic diesel engines with scrubbers are normally "greenest" for vessels with a range over >500 nm

4.2 LOGISTICALLY CONSTRAINED ASSETS

Based on analyses from the case study, car technology viewed as zero emissions will negatively impact vessels based on capability and or range requirements. Current energy density and fuel storage options need to improve for both battery and hydrogen for the negative effects of these power sources on a capability to be overcome. As a result overall vessel displacements are likely to increase if alternative green powering options are pursued.

Key findings in a logistical constrained asset are:

- Batteries are too heavy to be practically considered unless part of a hybrid solution.

- Hydrogen fuel storage is least affected by speed on short to medium duration tasking's. However, hydrogen storage is the biggest constraint on this solution.

- Synthetic diesel use is best suited for long ranges where large payload capacity is required. But compared to current solutions, scrubbers should be included to remove NOx emissions. The fitment of scrubbers will counteract some of the payload benefits of synthetic diesel compared to hydrogen.

5. REFERENCES

1. WIDDECOMBE, K., HEARD, A., WHITE, M., SHAMTOOB, R., PERES, G., FAROOQUE, I., defence Carbon baseline methodology and calculation. *Bristol : MoD*, 2021.

2. THE ROYAL SOCIETY. Sustainable synthetic carbon based fuels for transport: policy briefing. *London : The Royal Society*, 2019.

3. BLOUNT, D., Performance by design. *Virginia : Blount*, 2016.

4. NATIONAL ACADEMIES OF SCIENCES, ENGINEERING AND MEDICINE**.** Powering the U.S. Army of the Future. *Washington, DC : The National Academies Press,* 2021.

5. NAVAL TECHNICAL. Regulations For The Safe Storage And Handling Of Petroleum Products In Hm Ships And Submarines. *Bristol : Ministry Of Defence*, 2020.

6. BROOK, P S., COWEY, K., RYAN, A J., Future Fuels and energy Sources. *Farnborough : Qinetiq,* 2003.

7. LOVEGROVE, A., Barriers to the Implementation of Alternative Energy Sources. *s.l. : BAE Systems*, 2014.

8. CROLIUS, S H., Literature Review: Ammonia as a Fuel for Compression Ignition Engines. [Online] Ammonia Energy Association, 23 March 2020. [Cited: 22 July 2021.] https://www.ammoniaenergy.org/articles/review-of-ammonia-as-a-ci-fuel-published/.

9. UMAR, K., PRASHANT, B., GUPTA, R., NITESH. Hydrogen Fuel: Opportunities and Barriers. [Online] May 2009. [Cited: 30 July 2021.] https://www.researchgate.net/publication/234112121_Hydrogen_Fuel_Opportunities_and_Barriers.

10. YANMAR HOLDINGS CO., LTD. Yanmar Conducts Field Demonstration Test for Maritime Hydrogen Fuel Cell System. [Online] Yanmar, 24 March 2021. [Cited: 22 July 2021.] https://www.yanmar.com/global/news/2021/03/24/89229.html.

11. INTERNATIONAL COUNCIL OF COMBUSTION ENGINES. *CIMAC White Paper 2 - Zero and Net Zero Carbon Fuel Options.* 2020.

12. LIS, A., Interview with Dave Morrison, Ricardo Motorsport, on the Ricardo / Engine Developments Diesel Le Mans Engine Project. [Online] daily sports car, 2003. [Cited: 28 July 2021.] http://archive.dailysportscar.com/subscribers/technical/ricardodiesel.htm.

13. ROWEN, A L., Ship Design & Construction, Volume 1. [ed.] *Thomas Lamb. Alexandria : Society of Naval Architects and Marine Engineers (SNAME), 2010. Vol. I, ISBN 0-939773-40-6.*

14. Sulphur limit in ECAs - increased risk of PSC deficiencies and detentions. [Online] Det Norske Veritas & Germanischer Lloyd , 12 March 2019. [Cited: 12 March 2020.] https://www.dnvgl.com/news/sulphur-limit-in-ecas-increased-risk-of-psc-deficiencies-and-detentions-142911.

15. KARAN, C., 10 Technologies/Methods for Controlling NOx & SOx Emissions from Ships. [Online] Marine insight, 23 12 2020. [Cited: 28 07 28.] https://www.marineinsight.com/tech/10-technologiesmethods-for-controlling-nox-sox-emissions-from-ships/.

16. Department for transport. Government takes historic step towards net-zero with end of sale of new petrol and diesel cars by 2030. [Online] UK goverment, 18 November 2020. [Cited: 29 07 2021.] https://www.gov.uk/government/news/government-takes-historic-step-towards-net-zero-with-end-of-sale-of-new-petrol-and-diesel-cars-by-2030.

17. SHUNSUKE, T., China plans to phase out conventional gas-burning cars by 2035. [Online] 27 October 2020. [Cited: 22 July 2021.] https://asia.nikkei.com/Business/Automobiles/China-plans-to-phase-out-conventional-gas-burning-cars-by-2035#:~:text=China%20plans%20to%20phase%20out%20conventional%20gas%2Dburning%20cars%20by%202035,-Latest%20goals%20call&text=In%202019%2C%20new%2Denergy%.

18. BBC News. Bentley reveals plan to go fully electric by 2030. [Online] BBC News, 6 November 2020. [Cited: 29 July 2021.] https://www.bbc.co.uk/news/business-54835588.

19. CAMPBELL, P., Proposed ban on diesel HGV Sales 2040 demmed unworkable. *Financial times.* [Online] 17 April 2019. https://www.ft.com/content/910be246-6058-11e9-a27a-fdd51850994c.

20. MCINTOSH, J I L., How It Works: Cylinder deactivation. [Online] Driving, 14 Febuary 2018. [Cited: 29 July 2021.] https://driving.ca/auto-news/news/how-it-works-cylinder-deactivation.

21. US Department of Energy. DOE HTCD program. [Online] US Department of Energy, 03 2014. [Cited: 2 12 2020.] https://www.energy.gov/sites/prod/files/2014/03/f9/2006_%20deer_milam.pdf.

22. GROVES, J., Mazda Skyactiv-X engine explained: squeeze or spark. [Online] Car Magazine, 15 October 2019. [Cited: 30 November 2020.] https://www.carmagazine.co.uk/car-news/tech/mazdas-skyactiv-x-petrol-compression-ignition-engine-details/.

23. SCARBOROUGH, C., In-depth: the Formula E powertrain explained. [Online] Car Magazine, 30 July 2020. [Cited: 30 November 2020.] https://www.carmagazine.co.uk/electric/formula-e-powertrain/.

24. MORRIS, C., Elon Musk: Cooling, not power-to-weight ratio, is the challenge with AC induction motors. [Online] Chraged Eletric Vehicles Magazine, 4 November 2014. [Cited: 29 July 2021.] https://chargedevs.com/newswire/elon-musk-cooling-not-power-to-weight-ratio-is-the-challenge-with-ac-induction-motors/.

25. Green Car Congress. BMW's hybrid motor design seeks to deliver high efficiency and power density with lower rare earth use. [Online] Green Car Congress, 13 August 2013. [Cited: 29 July 2021.] https://www.greencarcongress.com/2013/08/bmw-20130812.html.

26. ROHM semiconductor. ROHM supplies Full SiC Power Modules to Formula E racing team Venturi. [Online] ROHM semiconductor, 30 November 2020. [Cited: 30 November 2020.] https://www.rohm.com/news-detail?news-title=rohm-supplies-full-sic-power-modules-to-formula-e-racing-team-venturi&defaultGroupId=false.

27. GE Aviation. https://www.geaviation.com/military/systems/silicon-carbide. [Online] GE Aviation, 30 November 2020. [Cited: 30 November 2020.] https://www.geaviation.com/military/systems/silicon-carbide.

28. ERICSEN, T., et al. Advances in SiC-based power conversion for shipboard electrical power systems. 2015 IEEE 3rd Workshop on Wide Bandgap Power Devices and Applications (WiPDA). *Blacksburg : IEEE*, 2015.

29. M FEDKIN, J A., DUTTON. 6.5. Efficiency of Inverters. [Online] Penn State Universtiy, 2020. [Cited: 16 August 2021.] https://www.e-education.psu.edu/eme812/node/738.

30. UK goverment. Government takes historic step towards net-zero with end of sale of new petrol and diesel cars by 2030. *UK goverment web site.* [Online] 18 November 2020. https://www.gov.uk/government/news/government-takes-historic-step-towards-net-zero-with-end-of-sale-of-new-petrol-and-diesel-cars-by-2030.

31. Defence Suppliers Forum: Research Technology & Innovation Group (RTIG). Powering Future Operations: Net Zero Challenges & Opportunities. [Online] December 2020. [Cited: 22 07 2021.] https://assets.publishing.service.gov.uk/government/uploads/system/uploads/attachment_data/file/967965/Energy_and_Power_Deep_Dive_Final.pdf.

32. THOMAS, C E., (Sandy). Fuel Cell and Battery Electric Vehicles Compared. *Alexandria, Virginia : H2Gen Innovations, Inc.*, 2009.

33. BILAL, A., All Cars In Norway Will Be 100% Electric By 2025. [Online] WCCFtech, 23 Febuary 2017. [Cited: 22 July 2021.] https://wccftech.com/norway-electric-cars/.

34. LAMBERT, F., The Netherlands reaches impressive 69% all-electric market share. [Online] Electrek, 8 January 2021. [Cited: 22 July 2021.] https://electrek.co/2021/01/08/the-netherlands-69-all-electric-market-share/.

35. International Maritime Organization. Initial IMO GHG Strategy. [Online] International Maritime Organization (IMO, 2019. [Cited: 17 August 2021.] https://www.imo.org/en/MediaCentre/HotTopics/Pages/Reducing-greenhouse-gas-emissions-from-ships.aspx.

36. SHRUBB, R., Hydrogen: the good the bad and the ugly. [Online] B2B connecting e-Mobility EV, 21 August 2018. [Cited: 22 July 2021.] https://www.b2bev.com/hydrogen-the-good-the-bad-and-the-ugly/.

37. Energy Observer. Appraisal of a transatlantic passage in solitude. *Energy Observer.* [Online] 7 May 2020. https://www.energy-observer.org/innovations/appraisal-transatlantic-passage-solitude.

Report documentation page

v5.0

* Denotes a mandatory field

1a.	**Report number: *** Report number	**1b.**	**Version number:** Version
2.	**Date of publication: *** Publication date	**3.**	**Number of pages:** 16
4a.	**Report UK protective marking: *** UK OFFICIAL		
4b.	**Report national caveats: *** NONE		
4c.	**Report descriptors: *** NONE		
5a.	**Title: *** Title		
5b.	**Title UK protective marking: *** UK OFFICIAL		
5c.	**Title national caveats: *** NONE		
5d.	**Title descriptor: *** NONE		
6.	**Authors: *** Rusne Ramonaite		
7a.	**Abstract: *** Abstract		
7b.	**Abstract UK protective marking: *** UK OFFICIAL		
7c.	**Abstract national caveats: *** NONE		
7d.	**Abstract descriptor: *** NONE		
8.	**Keywords:** Keywords		

Please note: Unclassified, Restricted and Confidential markings can only be used where the report is prepared on behalf of an international defence organisation and the appropriate prefix (e.g. NATO) included in the marking.

* Denotes a mandatory field

9. Name and address of publisher: * [Company] [Company Address]	**10. Name and address of funding source:** Specialist Systems Programme Manager Dstl Porton Down

11.	**Funding source contract:**	STECH/555	
12.	**Dstl project number:**	PJ100218	
13.	**Programme:**	Specialist Systems	
14.	**Other report numbers:**	Other report numbers.	
15a.	**Contract start date:**	01/04/2018	**15b. Contract end date:** 31/03/2022
16.	**IP conditions for report: ***	MOD funded Dstl authored report	
17a.	**Patents:**	NO	
17b.	**Application number:**	Application number	
18.	**Release authority role:**	Project Manager	

Guidance on completing the report documentation page can be found on the Gov.UK website.

AVOIDING ALUMINIUM HSC HULL DAMAGES DURING DRY DOCKING AND OCEAN TRANSPORT – OPERATOR'S PERSPECTIVE

D Tugay, University of Southampton, Turkiye

SUMMARY

The dry standing of aluminium vessels requires thorough consideration of the placement of support points across the hull due to certain material properties. Yards are required to precisely position the keel blocks at complete matching accuracy with stronger bulkhead frames, risking damage to the hull otherwise. This is not always possible due to external factors affecting the dry-docking procedure, as such a solution is required to increase the area suitable for block loading and make regions across the bulkhead frames stronger to tolerate misplacement. In this paper, a solution for reinforcement of the bulkhead frames and adjacent areas is proposed, while the original and improved geometries were compared using Finite Element Analysis (FEA) simulations. The results and other improvements are discussed, with the general preference from the operator noted.

1. INTRODUCTION

Dry docking is an essential part of operating a vessel due to the requirement of surveying, inspection, and maintenance for the craft. In high performance aluminium vessels, where minimising hull resistance is of high importance, frequent dry docking for the purpose of hull cleaning and antifouling reapplication may be required. Small aluminium craft are typically placed on slipway cradles or vertical lift systems for dry docking, as such systems are more compact and often more economical in comparison with floating or graving docks.

Due to the ductile nature of aluminium, loads are required to be distributed along the strongest areas in the hull. Loading points and their correspondence with required frames and bulkheads lie key in the proper preparation of the lifting system. For slipway dry docking, the support blocks on the cradle should precisely lie on bulkhead frames, as this guarantees the avoidance bending of shell plating. The same is required for travel lifts, however the problem arises when slings of the lift are placed on the very same bulkhead frames to be used for dry standing supports, as the keel blocks could not be used in the same location.

Although operators and yards aim for precise placement of the vessel on the cradle, there are instances where the process becomes extremely difficult due to external factors. When the load support points are misaligned even slightly and the main force is exerted on the ordinary frames, the risk of inward bending of the keel plates arises, leading to possible damage of the keel of the vessel. When bulkhead frames are occupied by lifting slings, the operator is forced to use other ordinary frames as loading points, leading to excessive areas inaccessible for maintenance. As a result of high stress concentration regions, the formed bends compromise the structural integrity of the vessel, disrupt the waterflow around the hull, negatively impact coatings applied to the hull, and effect the overall seaworthiness of the vessel.

This paper will discuss the main practical issues experienced by aluminium high speed craft operators during dry docking and ocean transports, outline processes implemented to avoid possible damage, and suggest an improvement in design preferred by the operators.

2. SLIPWAY CRADLES

2.1 BACKGROUND

The slipway system allows the transport of the vessel out of the water with the aid of a cradle moving on a sloped plane made of solid material. The system allows for transportation of craft of various size and is often preferred for catamarans due to their increased breadth and probability of compressive damage of the hull from the lifting slings. A typical cradle used in dry docking of medium size aluminium catamarans consists of two parallel located hollow steel bars attached together with four transverse steel bars by welding. Roller or wheel aided chassis is often used for transportation of the cradle, where wheel cradles offer the benefit of precise positioning due to their steering capability. A typical arrangement is represented in figure 1 below.

2.2 OPERATIONAL CHALLENGES

Due to weight considerations, the thickness of shell plating on medium size aluminium high speed passenger vessels does not often exceed 5-7mm, while stiffener and bulkhead thicknesses do not exceed 30-50mm. Although providing great performance due to low weight, the combination of higher ductility of aluminium in comparison with other alloys and reduced thickness of the plates poses several challenges during dry docking operations.

Because of abovementioned factors, the operators are forced to place the vessels precisely on bulkhead frames, and lie the keel blocks to have complete matching accuracy with the bulkhead positions of the vessel. For a typical medium size passenger high speed ferry subdivided into 7 compartments and 6 bulkheads, the number of areas that can be used as loading points becomes deficient. As shown in figure 1 below, during a typical dry dock of a 40m craft on a 30m long cradle, a minimum of 4 out of 6 possible bulkhead frames (note: the midship bulkhead cannot always be used as main support point due to cradle limitations) are required to be used as loading points. As a result, avoiding damage due to overloading of the shell plate at the lifting points and spreading the load across those support points becomes more difficult, especially due to often uneven weight distribution across the vessel.

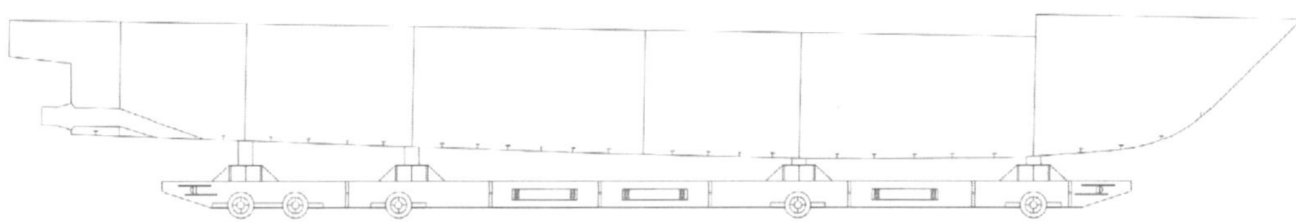

Figure 1: Hull of aluminium vessel on cradle - lifting arrangement.

In order to spread the weight, operators may need to include extra keel blocks below the ordinary frames to provide support across the hull, however, this also poses several challenges. Firstly, when additional keel blocks are installed on the cradle, the risk of damage due to misalignment of the blocks with the hull increases as the blocks are to lie on ordinary frames which cannot tolerate high loads. Secondly, in instances where the vessel is too densely supported on the hull as shown in figure 4, the additional keel block supports may restrict the flexibility of the works that can be carried in the shipyard (shell plate replacement, paint application, blasting). The location of appendages such as sacrificial anodes, sea chests, echo sounders, and other equipment can also make the installation of additional support points difficult. Finally, the areas of the hull occupied by the blocks become inaccessible for maintenance, survey, and paint works, potentially increasing the drag caused by the hull form.

Figure 2: Hull lifted with 4 lifting points.

Figure 3: Load spread across multiple lifting points.

When lifting using only bulkhead frames as loading points, high precision in placement of the vessel on the cradle becomes crucial. In this case, the weight of the vessel is concentrated only on the bulkhead frames, therefore the loading experienced around the bulkhead keel plates is higher. Even minor misalignment of the keel blocks with the designated loading points can cause the thin aluminium shell plating and stiffeners to bend, leading to hull damage.

Due to the slipway cradle applications being carried out on a sloped surface, it is impossible to place the vessel on all loading points simultaneously. If the bow of the vessel is being laid on the fore keel block at the design waterline first, the vessel is then held with the mooring lines as the cradle is operated up the slipway. This causes loss of buoyancy compensated by the fore block reaction force at the bow, leading to the vessel trimming by stern until the hull is sat on all keel blocks.
In practice this process poses several challenges, which makes it extremely difficult to accurately place the hull on the cradle.

2.2 (a) Position fixing on the cradle

It is difficult to fix the vessel along the cradle to prevent movement in longitudinal and transverse directions whilst the vessel is still buoyant. Mooring lines and straps become ineffective when attached to the cradle, as in most cases they can only be attached vertically or at steep angles. Additionally, the tension on such lines changes with the trim of the vessel, which can result in damage to the deck mooring equipment at the bow of the ship as loading increases with astern trim. Shore infrastructure may also not allow for placement of mooring lines and may still result in vessel swinging. It is also not practicable to hold the vessel with stoppers, as this can cause further damage.

2.2 (b) Underwater visibility

Divers may be required to confirm the correct position of keel blocks on the bulkheads during the docking. Poor underwater visibility caused by disturbance on of the seabed can result in divers having difficulties or even being unable to confirm the correct positioning of the cradle.

2.3 (c) Climate conditions

Changes in wind, tide, and waves at the time of the docking pose challenges in correct alignment of the craft. The lightweight and high surface area nature of aluminium passenger ferries can lead to wind induced motion. Precise placement may not be achieved as such changes may cause motion of the vessel during the docking.

2.4 (d) Wake from passing by vessels

Wake effects from passing traffic can negatively affect the accuracy of docking. The waves can lead to the vessel regaining buoyancy and potentially causing a shift in position relative to slings.

Even slight shifts of only 5-10 cm can cause the supports to lie off the bulkhead on either side, leading to weaker unsupported areas around the frame to take the loads. As no reinforcements around the bulkheads across the keel are usually present, the load is then only applied to the shell plating and ordinary stiffeners, often causing those elements to bend.

3. LIFTING BY SLING OPERATIONS

3.1 BACKGROUND

Aluminium high speed craft can also be lifted with the aid of slings attached to a lifting mechanism such as a crane on board a cargo vessel or designated travel lift at a shipyard. A travel lift is a specialised mobile crane, typically consisting of two rectangular side frames joined by a beam across one or both ends equipped with a hydraulic lifting mechanism and a set of adjustable slings. The slings are positioned beneath the vessel's hull, allowing hoisting and transportation to a designated area. Cranes on board cargo vessels are often used to lift craft from and off the deck of carriers for transportation purposes. Slings can be attached to the crane, consisting of a main structure, lifting wires, and spreader bars.

3.2 OPERATIONAL CHALLENGES

For craft not equipped with lifting lugs, the slings are positioned underneath the hull for lifting and launching purposes. Similar to cradle operations, the operator will seek to place the lifting slings on the strongest areas of the hull – along shell plates of bulkhead frames. This poses several challenges related to the placement of the slings and their operation.

3.2 (a) Travel Lift Operations

If a typical scenario of a vessel lifted with 4 slings by a travel lift is considered, there are a minimum of 4 lifting points required on each side of the hull, meaning that the slings will occupy 4 bulkheads. For a medium size aluminium ferry typically subdivided into 7 compartments and 6 bulkheads, the slings leave 4 out of 6 loading points unusable for the shipyard as the slings can not be placed on keel blocks due to the risk of compressive damage. The travel lift already utilises all usable space on the hull for support placement, leaving the shipyard with no choice but to use frames not designated for seating, which obliges them to use too many frames to make sure no damage occurs. As a result the vessel is too densely supported on the hull restricting the flexibility of access and the works that can be carried on the shipyard (shell plate replacement, paint application, blasting). This leaves the vessel with areas under support not accessible for painting, causing launching without antifouling protection and reducing performance.

3.2 (b) Heavy Lift Operations

For lifting on an off a heavy lift cargo carries using a crane, the limited amount of support points for craft without lifting lugs causes similar issues. Due to the lifting slings occupying the area required for keel block supports on the deck, the operator is again forced to lay the vessel on increased number of keel blocks, exerting the force from supports onto the hull at points not designed for support loading. In instances where the deck of the carrier has limited space, the crane operator may also require the lifting of the vessel from slightly different areas to be able to fit it on the deck. The restricted number of lifting points makes this operation extremely difficult.

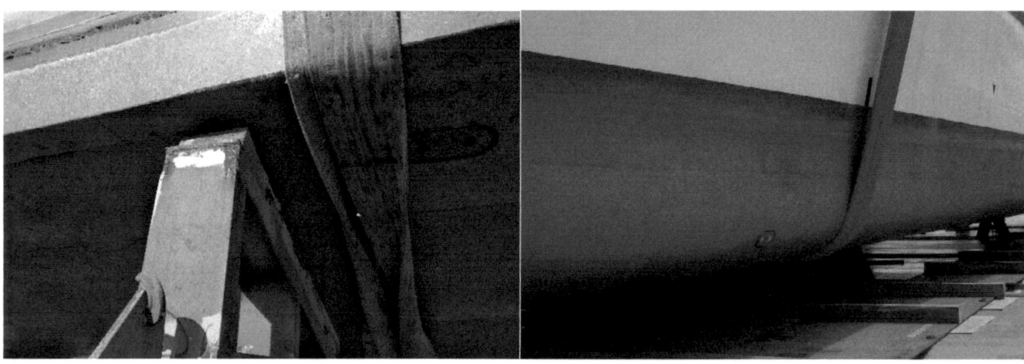

Figure 4: Placement of hull on bulkhead support point.

Figure 5: Load spread on ordinary frames.

4. SUGGESTIONS FROM OPERATOR

4.1 OPERATOR PREFERENCE

Operators would prefer more flexibility in terms of areas of the hull available for use as loading points during dry docking and transportation. For slipway applications, the importance of the hull to tolerate minor shifting during docking would prevent most damages that occur during the docking process. Reinforcement of the area between the bulkhead frame and the adjacent transverse frames would allow the previously unsupported shell plate between the frames to have greater resistance to bending. The reinforcement would help transfer the load from the shell plating to the frames, and hence reduce risk of indents and gashes on the hull.

Similarly, for lifting using slings, it would be preferred to have more areas that the slings can be laid and operated from. If a vessel is equipped with two adjacent frames that can be used as loading points, then one can be used for the sling, while the other can be used for keel block placement.

4.2 PROPOSED SOLUTION CONCEPT

Due to high emphasis on weight efficiency of high speed craft, the proposed solution is required to be designed to add minimum extra weight to the structure, while providing maximum support for the weak areas across the bulkhead frames to tolerate misplacement and allow distribution of the load for sling operations. The improvement is to also provide strengthened transverse areas to reduce the possibility of shell plate bending.

As a result, a rectangular geometry which is to extend longitudinally between the bulkhead and the adjacent frames is considered to allow more flexibility in dry docking operations for operators and yards.

4.3 STUDY GEOMETRY AND DIMENSIONS

To compare stress and deflection values for standard and improved versions of the hull, two geometries were created and load simulations were done with the aid of Finite Element Analysis (FEA). The typical cross section of a 40 metre high speed aluminium passenger catamaran at the midship region is shown in figure 6 below. The 3D model is of a 10m long midship section, where the hull inner structure consists of longitudinal stiffeners and transverse frames, including a bulkhead frame in the centre of the model. For the purposes of the study, the shell thickness is modelled to be equal to 6mm, bulkhead, as well as ordinary longitudinals and transverses to be 10mm thick. The longitudinals were modelled to extend 60mm inward of the hull with a simple bulb modelled as a filleted 20mm by 10mm rectangular geometry at the end to increase stiffness. A centre girder with thickness of 10mm and height of 50mm was modelled across the keel of the hull. The transverse frame spacing was set to 1000mm, while the longitudinals were spaced 250mm apart.

For the improved version, the hull was reinforced by a 10mm thick rectangular structural element, with a height of 30cm and the length of 100cm, which was placed to connect the adjacent frame with the bulkhead. The geometry also included a 10mm thick 80mm long transverse element attached to help spread the load concentrated at the shell plates, forming an inverted T-section. The triangular reinforcement was placed on 4 areas in between ordinary longitudinal frames, as shown in figure 7 below.

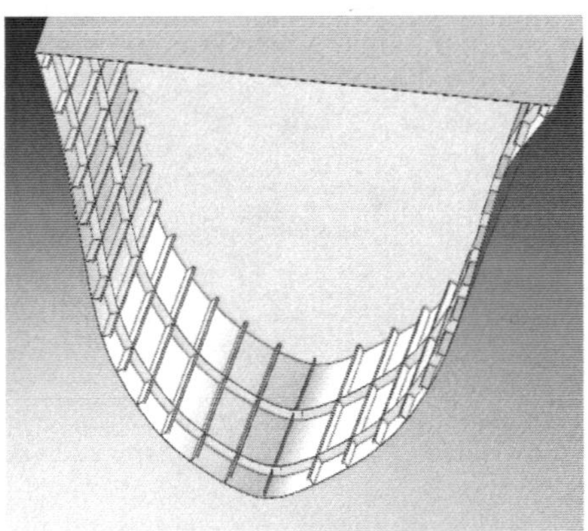

Figure 6: Original geometry.

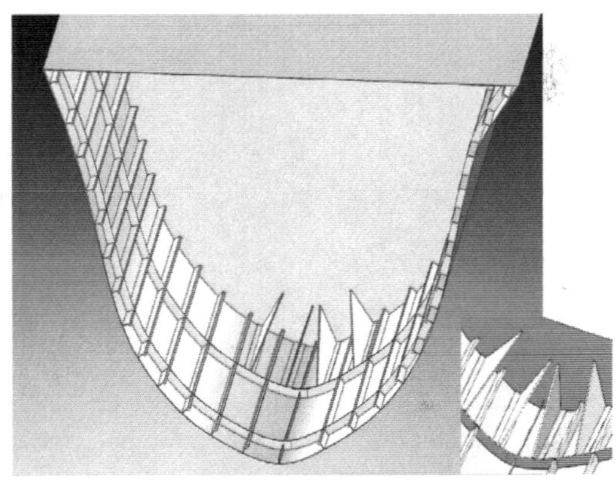

Figure 7: Modified geometry.

4.4 FINITE ELEMENT ANALYSIS

The Marine Grade Aluminium 5086 Alloy was used for the simulation. A concentrated force was applied to an area on the geometry of the hull projected from a 30cm by 50cm rectangle, which is used to approximate the real life area of the keel block support touching the hull. To account for the worst case scenario, the concentrated force was applied at a longitudinal distance of 30cm from the bulkhead. Four studies were conducted with four different magnitudes of loading. The typical loading for a support of 40m high speed passenger catamaran was calculated assuming the weight of the craft to be 150 tonnes, and the craft to be supported by 4 keel blocks on each hull. The average total force exerted on each support point in Newtons was calculated as shown:

$$Weight\ on\ support\ point\ [kg] = \frac{150\ 000}{8} = 18750\ kg$$

$$Force\ on\ support\ point\ [Newtons] = w * g = 18750 * 9.81 = 183937.5\ Newtons$$

The average force on each support point was then rounded to 200kN to simplify the calculation. To investigate the behaviour of the hull in different loading scenarios, the analysis was done with three other loadings, one force value 50% below the rounded average, and two forces above the average by 50% and 100% respectively were considered. The following forces were considered: 100kN, 200kN, 300kN and 400kN. The geometry was fixed from the deck and the edges of the hull, which was assumed satisfactory to approximate real life loading scenario.

Table 1 below shows the maximum Von Mises [N/m^2] and Deflection [mm] values for the 4 forces applied at the two geometries (original and reinforced).

Table 1. Comparative results of Von Mises Stresses and Displacement for original and modified geometry for a range of forces applied.

	Original		Modified	
Force, N	Von Mises, N/m^2	Deflection, mm	Von Mises, N/m^2	Deflection, mm
100 000	3.986e+7	0.501	7.424e+6	0.293
200 000	5.816e+7	1.078	1.609e+7	0.623
300 000	9.517e+7	1.542	2.069e+7	0.958
400 000	1.012e+8	1.904	4.016e+7	1.031

Figure 8 and 9 represent the resultant displacement graphs for the original and modified geometry, where the difference between the plate displacement magnitude and location is clearly seen.

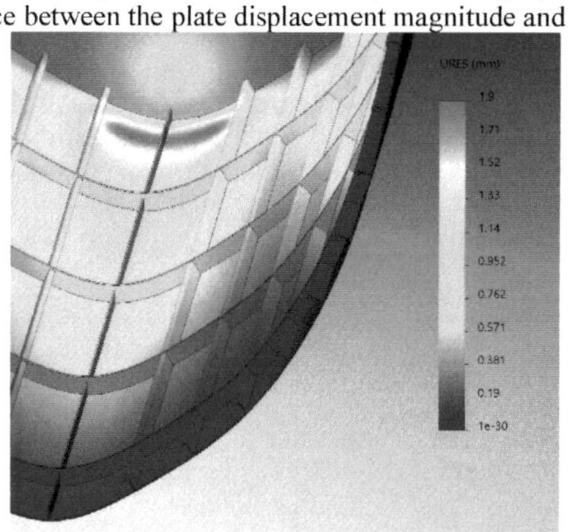

Figure 8: Original geometry FEA at F=400kN.

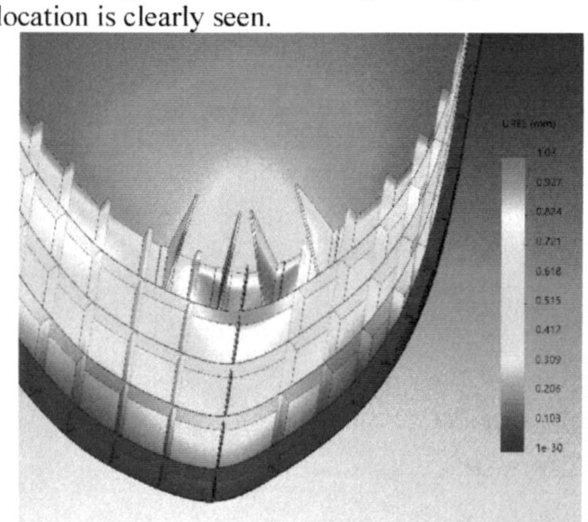

Figure 9: Modified geometry FEA at F=400kN.

As can be seen from table 1, both the Von Mises and the Displacement values vary significantly between the results from the original and improved geometry. As expected, both the Von Mises stresses and the Displacement increase with the increase in the applied force magnitude. When the force of 100kN is applied, the Von Mises stresses are 3.986e+7 N/m^2 and 7.424e+6 N/m^2 for original and modified geometry respectively, showing a percentage difference of 138%. The resultant displacement of the geometry follows a similar pattern, where for the abovementioned force, with values 0.501mm and 0.293mm respectively, or 53% reduction for the improved geometry. For the maximum applied force of 400kN, the Von Mises stresses reach 1.012e+8 N/m^2 and 4.016e+7 N/m^2 for the original and reinforced geometry respectively, giving the percentage difference of 86%. The displacement values for the same force were 2.038mm and 1.031mm for the original and reinforced geometries, with the displacement for the original geometry 60% higher than for modified. It should be noted that although the Von Mises stresses do not reach the Yield Strength for the material which is equal to 2.071e+8 N/m^2 [1], it is possible that for worse loading scenarios (such as smaller support areas of the blocks or force concentrated across sharper keel geometry at the bow) the Von Mises can exceed the Yield Strength and cause permanent deformation.

Figures 8 and 9 allow for a graphical comparison between the two designs to be made. In the modelled event of vessel position shifting 30cm forwards from the required keel block location, the displacement of the shell plates and structural elements is shown. As can clearly be seen, in the original design the force is concentrated across the sitting area of the keel block and the centre girder. Due to the lack of structure designed to spread the load, the displacement is high and this loading experienced in real life could have potentially resulted in hull damage. In the case of the geometry modified with internal stiffening structure as shown in figure 8, the maximum displacement distribution reduces dramatically. The previously unsupported areas around the bulkhead frames now include protective rectangular plates, designed to transfer the misplaced load to the bulkhead. The displacement concentration areas have now shifted from the shell plate to the reinforcements, with the maximum displacement approximately 60% lower in magnitude, which is an expected and desired result.

4.5 DISCUSSION

The improvements of the hull geometry were suggested based on the real life issues witnessed by the author, with the aim to improve the loading tolerance of the hull across bulkhead frames. The suggested improved geometry is only a concept developed to illustrate possible issues which occur during operation of the craft and is presented with emphasis to visualise the problem and potential solution. It is important to note that the geometry suggested for this study was created only for demonstration purposes of one possible solution to the problem, and other improvements can be made to achieve better results.
Overall, the aim of the study is to demonstrate that yards and aluminium high speed craft operators would prefer to have more reinforced areas across the hull to be designed as loading points. Shipyards considering additional supports across the bulkhead frames would allow for greater flexibility of operation during dry docking using both the cradle and the sling arrangement.

5. CONCLUSIONS

This paper looked at challenges faced by operators during the dry standing of aluminium high speed vessels. The nature of challenges and their cause was investigated for slipway and sling lifting operations, while differences between two approaches discussed. For lifting using slipway cradles, main issues occur due to the possibility of hull damage in case the supports are misaligned with bulkhead frames. For sling lifting operations, the reduced number of areas on the hull that can be used as support points causes difficulties for the operators when placing the vessel from the slings to the keel block supports. For both cases, operators would prefer to have more areas across the hull designated as lifting points, as this would allow for flexibility in dry docking operations. To achieve that, an FEA study was conducted, comparing loading experienced on a typical midship hull geometry versus a modified version with reinforcements. It was determined that the reinforcements on the improved geometry allowed both the Von Mises and Displacement values to reduce significantly. This proves the effectiveness of the improvement, and shows that such structural elements can help prevent hull damage due to minor misplacements in the future. Operators would be grateful if shipyards were to expand areas designed as loading points, allowing the hull to tolerate minor load shifting.

6. REFERENCES

1. STRUCTURAL ALLOYS HANDBOOK, 1996 edition, John M. (Tim) Holt, Technical Ed; C. Y. Ho, Ed., CINDAS/Purdue University, West Lafayette, IN, 1996.

7. AUTHORS BIOGRAPHY

Deniz Tugay is currently studying Naval Architecture and Marine Engineering at the University of Southampton in the UK. The author has also been involved in preparation and planning of annual maintenance and dry docking of aluminium passenger ferries. Working alongside the dockmasters the author assisted and documented dry dock maintenance work such as; cradle preparation works, hull repairs and coating application.

A LOITERING OPTIONALLY CREWED HIGH SPEED RESCUE VESSEL FOR REMOTE REGION RESCUE RESPONSE

E Jonze, Docksta Varvet (SAAB), Docksta, Sweden, **J Dalziel** and **R Pelot,** Dalhousie University, Halifax, Canada

SUMMARY

In the cold Atlantic waters north of Newfoundland, up through the Davis Strait into Baffin Bay (between Baffin Island and Greenland) there are vast areas with many navigational hazards, including heavy weather, freezing temperatures, sea ice. There is a limited amount of maritime traffic, however, ships (fishing trawlers, cargo ships, tankers, passenger liners) as well as smaller craft (local fishing vessels, yachts and local small craft) ply these waters. Inevitably incidents happen, with potentially tragic consequences. How best to be able to respond and minimize loss of life? There are few shoreside resources – air bases and ship bases are far away to the south. Only a limited number, if any, government ships are in the area, ships of opportunity may be able to help, helicopters are far way and long-range aircraft may be able to deploy pumps, life rafts and even SAR Techs.

One solution to speeding up response could be stationing on site a high-speed, uncrewed (optionally crewed) seagoing vessel. This vessel could be strategically located remote from other sea rescue resources but near areas of potential incidents. This presentation proposes a novel solution to the problem of rescue response. Current status of rescue response and marine incidents is analyzed, clarifying the need for rescue resources. The application of a Loitering, Optionally Crewed, High Speed Rescue Vessel will be evaluated, considering the many technical and operational issues involved. In addition, the issues of integrating this vessel with other SAR assets, to improve maritime safety, will be considered.

NOMENCLATURE

AIS	Automatic Identification System
CBC	Canadian Broadcasting Corporation, Canada's national television and radio network
CCC	Central Control Centre, term used in this paper for the primary remote control station
CCG	Canadian Coast Guard
DND	(Canadian) Department of National Defence
EO/IR	Electro-optical / Infrared Camera
IMO	International Maritime Organization
JRCC	(Canadian) Joint Rescue Coordination Centre
LHSRV	Loitering Optionally Crewed High Speed Rescue Vessel
MRCC	Maritime Rescue Coordination Centre
PTZ	Pan/Tilt/Zoom EO/OR Camera
SAR Tech	Search and Rescue technician, a specially trained member of an elite group of primary care paramedics that provide on-scene medical aid and evacuation, may be deployed by parachute
SISAR	Canadian Government Search and Rescue Program Information Management System (incident database)
TSB	(Canadian) Transportation Safety Board
UAV	Unmanned Aerial Vehicles
UMV	Unmanned Marine Vehicles (generally referring to smaller units)

1. INTRODUCTION

1.1 PREAMBLE - FROM CONCEPT TO REALITY: THINK SLOW, ACT FAST [1]

Over the past decade, there have been several concepts of how uncrewed vessels (aka Unmanned Marine Vehicles (UMV), referring to smaller craft) could contribute to maritime safety. If so, questions arise about: how would they help? what roles could they fulfill? what could they do? what could they not do? how would they fit into a maritime search and rescue system? [2], [3], [4].

In general, successful applications of UMV for SAR (Search and Rescue) have been for small craft, often for beach rescue or person-in-the-water response [5], or even proposed as air deployed devices [6].

However, not all the past efforts have been successful: sometimes they have met with resistance; perhaps insufficient study was done; the application was hurried; front-line operators were not consulted or involved; or it was handed over as a 'turnkey' device with insufficient training as to operation and reasonable expectations.

Successful implementation of a project, such as this, requires careful consideration by experienced, competent personnel at all levels: careful definition of needs and requirements, technical and regulatory compliance, and finally, front-line operations. Thus, the maxim to 'Think Slow (i.e., carefully), Act Quickly' [1] applies well in this case, as well as starting with small building blocks – one or two craft – to iron out the inevitable problems.

1.2 OVERVIEW

The International Convention for Safety of Life at Sea (SOLAS) was introduced largely due to the Titanic disaster. SOLAS Chapter V Regulation 7 states:

> *"Each Contracting Government undertakes to ensure that necessary arrangements are made for distress communication and co-ordination in their area of responsibility and for the rescue of persons in distress at sea around its coasts. These arrangements shall include the establishment, operation and maintenance of such search and rescue facilities as are deemed practicable and necessary, having regard to the density of the seagoing traffic and the navigational dangers and shall, so far as possible, provide adequate means of locating and rescuing such persons."*

The International Maritime Organization has continued to update SOLAS to improve safety. In addition, the International Convention on Maritime Search and Rescue (the SAR Convention), which entered into force in 1985, developed an international SAR plan, with nations being deemed responsible for Search and Rescue Regions (SRRs) so that the rescue of persons in distress anywhere at sea will be coordinated by a national SAR organization and, when necessary, by cooperation between neighbouring SAR organizations [7], [8].

However, some of these SRRs are very large, may be subject to heavy weather conditions and, in addition, may be far from bases for rescue resources. Hence, what is the best way to provide the necessary resources to save lives at sea, particularly in remote locations far from shore bases? Ships are expensive to maintain on station, relatively slow and impacted by heavy weather; fixed wing aircraft have limited time on station and limited rescue capabilities; and helicopters have limited range and time on scene and limited survivor capacity.

The advent of remotely operated and autonomous vehicles in recent years has brought this rapidly developing technology to bear on maritime search and rescue. Numerous devices have cropped up, aimed at diverse segments of the problem: near shore rescue, sets of drones to assist in wide-area searches, self-directed contraptions that could be launched from ships to assist with man overboard. Various overviews and reviews have been published to help capture this rapidly changing field. Jain et al. [9] produced a comprehensive description of various features of autonomous robotic boats for SAR based on a 5-year review, focusing particularly on navigation, propulsion and victim detection advances. Other publications [10] provide an overview of the key UMV products under development and their comparative capabilities.

One solution for Remote Region Rescue Response could be stationing on site a high-speed, uncrewed (optionally crewed) seagoing vessel.

- This vessel could be strategically located near areas of potential incidents that are otherwise remote from other sea rescue resources. It would require endurance to maintain its location on station, while reserving sufficient fuel to proceed at high speed to the site of a serious incident, remain on site for the required length of time to effect rescue, and then proceed to a safe place of refuge with the survivors.

- As an uncrewed vessel, its speed of response would not be constrained by health and safety considerations of crew members. The risk of injury, showing up immediately, or perhaps years later as a result of repetitive wave impacts, is a constraint faced by other high-speed rescue vessels, potentially limiting their speed of response. [5]

This paper illustrates the nature of the demand by displaying remote maritime incidents around Canada, and some examples of near tragedies in the Northwest Atlantic and then proposes a novel solution to the problem of rescue response. Current status of rescue response and marine incidents is assessed clarifying the need for rescue resources. The application of the Docksta Loitering, Optionally Crewed, High Speed Rescue Vessel will be evaluated, considering the many technical and operational issues involved. In addition, the issues of integrating this vessel with other SAR assets, to improve maritime safety, will be considered.

This paper is in conjunction with presentations given at the World Maritime Rescue Congress 2023 (Poster Presentation) and at the Royal Institution of Naval Architects Surv11, Surveillance, Search and Rescue Craft Conference, Rotterdam, June 18-22, 2023.

2. THE CURRENT SITUATION

Figures 1a and 1b: F.V. Atlantic Charger sinking near Resolution Island, southeast of Baffin Island, September 21, 2015. Rescue plane from Nova Scotia flew for 3.5 hours to arrive on scene. The crew spent 12 hours in a raft, were eventually saved by a vessel of opportunity, after a near tragedy with a first rescue attempt. (DND photos) [11] [12] [13]

After the fishing vessel Atlantic Charger sank in Davis Strait near Baffin Island, the lack of Canadian maritime rescue resources in this region was again raised.

> *The owner of the Atlantic Charger, which sank in the Davis Strait last month, is asking why a search and rescue helicopter wasn't dispatched from Labrador to rescue the crew. "They wanna know where the chopper was. They asked me, 'Where was the damn chopper,' and I started asking questions," Watkins told CBC's The Broadcast* [11]

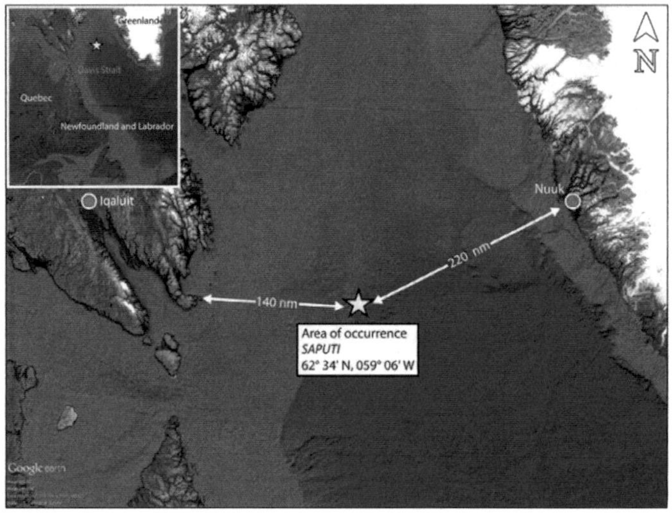

Figure 2: Shrimp Trawler Saputi incident [1]. Five months earlier, the Atlantic Charger sank about 140 nm west-southwest of this. In both cases Canadian air support was over 3 hours away, and no Canadian helicopters or rescue ships were readily available. (Map – TSB Canada) [14]

Five months later, there was another an example of a potentially disastrous incident in the Northwest Atlantic: in February 2016, the heavily ice-reinforced shrimp trawler Saputi, while towing her nets at 3 knots, collided with a growler (a piece of multi-year Arctic ice), and suffered hull damage leading to extensive flooding (Figure 2). Although a Canadian aircraft from Nova Scotia (about 1,200 NM away) was able to drop emergency pumps, and a Danish naval vessel was able escort her to Greenland, the Danish vessel's Rigid Hull Inflatable Boat was damaged and became unserviceable; had the Saputi, with a 45-degree list, capsized the 29 persons onboard would have been tossed into the frigid February, Arctic Ocean. |

> *"Nunavut (Canadian Arctic) officials press for an Arctic search and rescue base: 'If we have a major incident up here, we are in a very bad position to be able to respond ... The Danish Navy had to be called in for help because it would have taken the Canadian Coast Guard about two days to reach the ship .... In the first case last fall, I think it took about 10 to 11 hours to get search and rescue facilities in the area". Nunavut (northern Canada) official speaking to the Canadian Broadcasting Corporation* [15]

Many marine activities worldwide operate in relatively remote waters, whether shipping, fishing, passenger vessels, or occasional recreational boating. One such region, illustrative of the need for significant remote response capability, comprises the waters surrounding Canada. As an example of the demand for remote unmanned vessels, Figure 3 shows instances from the Canadian Coast Guard SISAR incident database that are relatively remote, occurring either 100 NM or more from shore, or any located in the north. These 280 incidents over 5 years are categorized by Distress (M1 in red), Potential Distress (M2 in orange) and non-Distress (M3 in green).

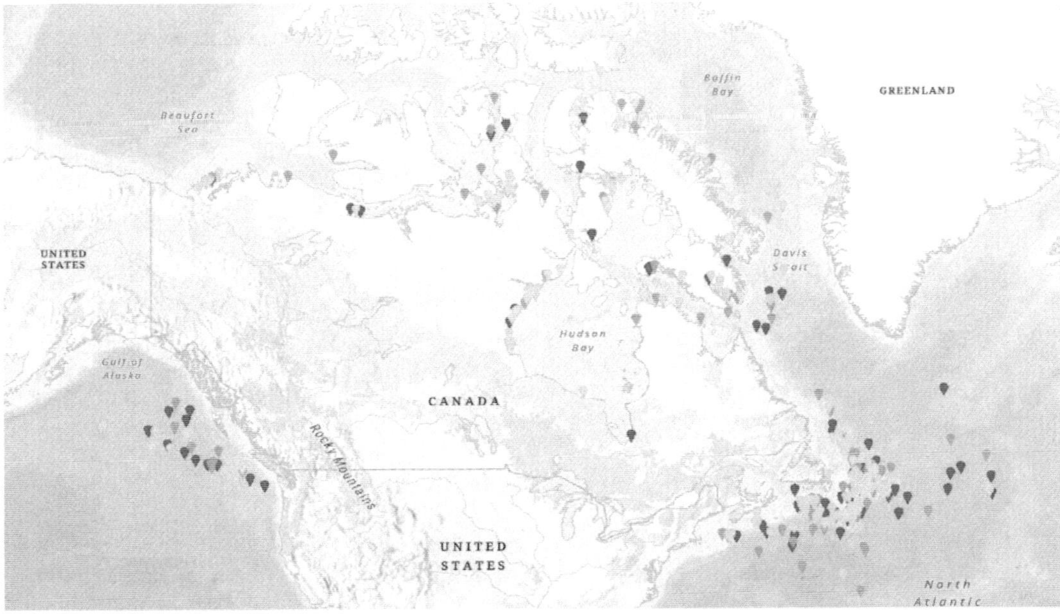

Figure 3: Maritime incidents in Canadian SRRs (100 nm from shore or in the north, 2016-20, (Dalhousie University, based on CCG SISAR database)

3. CONCEPT OF OPERATIONS

The intended use of the Docksta Loitering, Optionally Crewed, High Speed Rescue Vessel (LHSRV) is to act as an uncrewed vessel which can be stationed offshore, remain on station until needed and then proceed at high speed to the emergency incident site. It can then render assistance to the survivors, either as a vessel they can board, or standing by while they make their way to port or await rescue, by marshalling life rafts until rescue, working in conjunction with ships, aircraft and helicopters, acting as an on-scene communications hub. The vessel could optionally carry a crew, either rescue personnel who board in port or at sea, or perhaps SAR Techs who deploy from aircraft. These personnel could assist survivors to board, provide first-aid, and use the vessel as an on-scene control centre.

The vessel would require suitable facilities for survivors, giving them life-saving protection, until they could be transferred to shore, nearby ships or to helicopters. In particular, the ability to work with helicopters, utilizing its helicopter winching area, could allow rescue personnel to be deployed onboard, and for survivors to be lifted off.

The vessel would require onboard, remote (nearby and long-distance) and autonomous / uncrewed control and communication capabilities. It would also require onboard capabilities:

- to monitor navigational hazards,
- ensure safe operation considering weather conditions, and
- to enable identification and rescue of survivors.

It also requires the ability to work with other Search and Rescue resources both to aid search, and to effect rescue. These could be ships, aircraft, remotely based UAVs (Unmanned Aerial Vehicles - drones), and shore-based resources. It would require to be able to proceed at speed to the incident site, up to 30 – 40 knots depending on conditions (as an uncrewed vessel, the limiting factor would not be the health and safety of the crew, but only the ability of the vessel and its equipment to safely operate and survive).

3.1 A RESCUE RESPONSE SYSTEM

A Rescue Response System is envisaged. It would include one or more vessels of 15-20 meters, each with a rescue capacity of about 20+ survivors, covering a wide expanse of remote regions (possibly several remote regions) with the ability to stay on station for about 4 or 5 weeks (while retaining sufficient fuel to participate in a rescue).

Figure 4: Concept of Operations; while in 'Station Mode', several vessels can be controlled from one remote station via satellite link; in 'Rescue Response Mode' individual control shifts to a specific remote vessel or more localized station. Vessel layout, size, personnel retrieval systems and other characteristics may vary, within a set of parameters. Several designs could operate within individual fleets, to best suit rescue response needs.

In principle, the Rapid Response System would:

- Monitor from one up to ten vessels in 'Loitering' mode from a Central Control Centre (remote-control station)
- Work with one or more Maritime Rescue Coordination Centres (MRCC)
  - Report to, and if necessary, transfer control to the relevant MRCC, for instance,
    - Should a distress signal be received
    - Should the LHSRV (Loitering Optionally Crewed High Speed Rescue Vessel) detect indications requiring investigation or response
    - Should there be a communications link breakdown
    - Any other reason as determined by the MRCC or National Authority
  - Surveillance data could be monitored by the National Authority, in parallel with the Central Control Centre.
    - As in most cases the LHSRVs would be operated by National Coast Guards, which have other responsibilities in additional to Search and Rescue, some (non-navigational) surveillance feeds could go directly to the National Authority, bypassing the Central Control Centre.
  - If agreeable to the National Authorities, the cost of the Rapid Response System could be divided amongst several Nations, allowing greater financial feasibility in providing Rescue Response Resources in Remote Regions.
- The system could be scaled up or down to suit needs and available funds.

The operational profile could be:

- Proceed at moderate, fuel-efficient speed, from base port to its 'station' (alternatively may be towed).
- Loiter on station for required length of time, for example, 30 days.
- Return at moderate, fuel-efficient speed, from 'station' to base port (alternatively may be towed).
- Remain in base port for maintenance until being redeployed.
- If emergency response required, proceed at speed* to incident site, remain on site to render assistance as required, proceed at reasonable speed to port of refuge, or if necessary, provide temporary refuge while loitering until more assistance arrives.

- Speed would vary from full speed, about 40 knots in good conditions, downwards depending on weather conditions. Speed in the 'uncrewed' situation would likely be at least 50% higher than if personnel were onboard, and at least equal to that of a much larger surface vessel.
- Assuming three stations in the example of northeast Canada (say from Northeast Newfoundland to Baffin Bay, four vessels should be able to maintain full time coverage. As another example, assuming International Cooperation, six stations (and eight vessels) would cover much of the region from North Norway to Eastern Canada.

The operating system consists of:

- the Central Control Centre (remote-control station), for controlling the vessels;
- regional, local, or even portable (wearable) control stations;
- the vessels;
- A satellite-based redundant communication link would be required between the remote station(s) and the vessel(s).

The control of the vessel in all modes must be at least equivalent to having a crew onboard.

The vessel(s) have the following operation modes:

- Manual (onboard control)
- Autonomous station keeping
- Reconnaissance
- High speed transportation to rescue/emergency site
- On-scene rescue/emergency/communications hub operation
- Life support (of rescued personnel)
- Interaction with other SAR assets
- Transportation of survivors to a safe area
- Being towed
- Not under command / drifting mode (if control communication links are broken).

The system could also incorporate other search devices, such as drones, and be designed to interface, under some levels of control, with other SAR assets, such as ships, long range drones, aircraft, satellites. The system could, and should, start small. Perhaps working with an existing prototype, such as the Docksta Enforcer III, so that SAR personnel could evaluate and comment on working with a remotely controlled / autonomous vessel, suggest modifications, and determine whether, and how, to pursue this project. Admittedly, the prototype vessel would lack many of the features and capabilities of the full-fledged vessel proposed below, but it could give some insight prior to embarking on a more expensive full-scale trial.

Figure 5: A Rescue Scenario: Rendering firefighting assistance, potentially in a hazardous atmosphere. Rescuing personnel from liferafts. Working with other SAR assets, such as a helicopter.

4. LHSRV TECHNICAL STUDIES

4.1 PROTOTYPE - THE CB90 AND ENFORCER III

The Docksta Loitering Optionally Crewed, High Speed Rescue Vessel (LHSRV) is based on the Docksta Combat Boat CB90. With more than 250 boats operating worldwide, CB 90 has been a success story since the first boat was commissioned. A new model of the proven concept has been launched: CB90 Next Generation. It combines rugged aluminium construction, high speed performance, sea keeping and manoeuvrability, advanced electronics, yet straightforward functional design.

Principal characteristics of the CB90 are:

- LOA 16.3 m
- Beam 3.8 m
- Displacement, light 18.3 tons, maximum 24.5 tons
- Power 2 x 900 hp
- Speed, maximum over 40 knots, cruise 38 knots (sea state 1)
- Range at cruise speed 300 nm [16]

Figure 6: Docksta CB90 HSM

Figure 7: Swedish Navy Tests Saab Enforcer III Unmanned Sea Vessel (USV) During Exercise.

More recently, the autonomous / remotely controlled Enforcer III version has been tested with the Swedish Navy. After the trials, the Navy pointed out: "A risk assessment is made every time you send out a boat, and today we always have people onboard. But we see that it is now possible to send unmanned boats out in higher risk scenarios..." [17]

The CB90 and the remote control / autonomous system on the Enforcer III have provided Docksta / SAAB with a unique experience and capability to design, build, work with front-line personnel to successfully put into service this concept, and design and evaluate the Loitering Optionally Crewed High Speed Rescue Vessel search and rescue system.

4.2 LHSRV STUDIES TO DATE

Starting in early 2022 Dalhousie University and Docksta Varvet have explored the concept of a Loitering Optionally Crewed High Speed Rescue Vessel (LHSRV) for Remote Region Response. More detailed study was sponsored by VINNOVA, the Swedish Innovation Agency.

The project actually started much earlier. Since 2016 Dalhousie University has investigated the use of remotely controlled 'marine vehicles' to maritime search and rescue, and has made several presentations, such as to SURV 9 Conference, RINA Autonomous Ships Conference, the International Maritime Rescue Federation Future Technology Panel, WMRC2019, and others [2] [3] [6]. Docksta Varvet's background is outlined in the previous section.

4.3 LHSRV DESIGN

Remote controlled, or optionally crewed, the vessel is designed to improve maritime search and rescue capabilities. Its innovative design, operational flexibility, and advanced technology come from the extensive experience with the CB90 Class and autonomous / remotely controlled Enforcer III variant. This design, construction and operational experience will ensure the LHSRV can conduct rescue operations with exceptional speed, precision, and efficiency.

Constructed from aluminium, the LHSRV features a self-righting design that ensures the vessel can quickly recover from capsizes, enhancing its safety and survivability in rough seas. The superstructure is designed to minimize ice buildup, making the LHSRV suitable for operations in cold water environments. The hull, based on the successful Docksta CB90 design, is modified and further strengthened to withstand the tremendous forces encountered during high speed operation in rough seas.

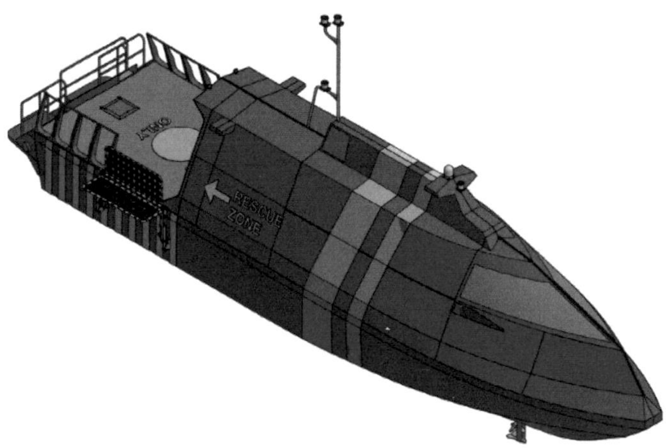

Figure 8: Naval architect's conceptualization of LHSRV. An alternative concept is shown in Figure 4.

The LHSRV's cabin is designed to meet the needs of a variety of rescue situations and includes the following compartments:

Emergency "Helmsman/navigator" seat: This compartment accommodates one person and is equipped with a simplified human-machine interface (HMI) for manual vessel control, allowing even those without prior knowledge to manoeuvre the LHSRV effectively. The compartment can be opened remotely or automatically in the event of a communication link loss.Bow thruster compartment: watertight compartment and access for maintenance / repair.

Life support compartments: These compartments can accommodate a total of 20+ survivors, excluding the helmsman/navigator seat. Designed with safety in mind, the spaces feature soft edges and corners to reduce the risk of injury. Each compartment is equipped with 20 seats, enabling survivors to adopt a half-lying position and fasten seatbelts for safety. The seats and floor position also feature straps to secure individuals in case of a roll-over. The compartments are connected to the remote station via video and intercom systems.

Equipment room: Contains the navigation/USV/electrical hardware. It can be placed aft of amidships to minimize the acceleration forces on the electronics, reducing the risk of electronics failure.

Storeroom: Contains lifesaving equipment, extra survivor suits, extra blankets, food, water, flares, and so on.

Engine room: Propulsion system, including any generators for the bow thruster /or pod system.

Fuel Tanks – Capacity about 10,000 liters, to meet operational requirements. Designed and located to suit vessel trim and stability requirements. Design to minimize risks of fuel contamination, and of pollution (should the vessel sustain damage). Arranged to suit easy and safe refuelling, including at sea or in remote ports.

Fore and Aft Peak Tanks, Void Spaces and Other Compartmentation – Void spaces and other watertight void compartments, watertight storage spaces for protection against flooding in case of damage and for damage stability.

4.4 MACHINERY INSTALLATION

The LHSRV is equipped with a retractable electrical pod forward, dual waterjets driven by V12 engines, a substantial battery pack, and a battery charging generator. enabling speeds well above 40 knots.

The electrical pod is used for precise station-keeping during loitering, while maintaining the vessel's head into the weather. The pod's retractable design minimizes resistance while travelling at high speeds, the risk of damage from ice impacts and improves the vessel's overall structural integrity.

The main engine / waterjet propulsion system enables high speed operation up to or over 40 knots, while also allowing controlled operation at low speeds. The waterjets minimize the hazards to persons in the water. Propulsion machinery installation designed to allow fast engine start-up when responding to a distress call, high speed, extended low speed operation, as well as minimizing the impact of fouling, slush ice, etc. The main machinery and generator are designed to withstand a rollover.

The generator / battery installation minimizes fuel consumption while 'loitering' on station. The power supply for essential services is designed to maintain these should the main batteries go flat, and the generator fail.

4.5 PERSON IN THE WATER RETRIEVAL SYSTEM

The Docksta designed person in the water retrieval system is depicted in Figures 9 and 10. When in situ it becomes a part of the railing. It could be operated locally or remotely. An incapacitated person could be picked up and deposited on deck (if no one is onboard it will be up to them to get into the cabin.

The proposed designs also include a ladder-like boarding area on the transom.

Alternative means of retrieving persons from the water could be considered.

Figures 9 & 10: Docksta person in the water (man overboard) retrieval system

4.6 HELICOPTER OPERATIONS

A helicopter lift area is marked on the main deck. Direct access is provided into the cabin, for both persons and stretchers. Personnel could also land in, or be hoisted from, the water.

4.7 COMMAND AND CONTROL SYSTEM

Remote control capabilities enable a single Central Control Centre (CCC) to control and manage multiple LHSRVs, providing the same or better situational awareness as an onboard navigator. The LHSRV utilizes dual redundant, independent satellite communication systems, ensuring reliable and consistent communication between the vessel and the CCC.

The LHSRV features several operational modes activated from the CCC (remote station), including manual mode, station mode, reconnaissance mode, high-speed transportation to the rescue/emergency area, in-area rescue/emergency operation, life support, transportation to a safe area, on/offboarding mode, towing mode, refuel mode, and NUC/drifting mode. These diverse operational modes allow the LHSRV to adapt to various situations and environments, ensuring effective search and rescue capabilities.

Advanced onboard AI technology assists in identifying and tracking objects of interest in the LHSRV's vicinity, enhancing situational awareness, improving search efficiency, automating object detection, and providing integrated decision support. This AI system allows the vessel to optimize search patterns and prioritize rescue targets effectively.

4.8 SEARCHING AND NAVIGATIONAL SAFETY CONSIDERATIONS

The requirements for searching, and for safe navigation, overlap in many regards.

The vessel is equipped with:

- EO/IR (electro-optical / infrared) cameras fixed or with a PTZ (pan-tilt-zoom) capability,
- radar
- searchlights
- bright deck lighting, including in the sea alongside the vessel

which allow the operator excellent visibility in most weather conditions.

The vessel has navigation lights and sound signals in compliance with the Collision Regulations, has AIS, transmitting its location and track, and receiving that of nearby vessels, and automated warning regarding radar and camera targets received.

While in the Autonomous Mode, if nearby targets are received which pose either a potential danger or a potential casualty, a message would be sent to the Central Control Centre, and possibly the relevant MRCC, informing them of the developing situation and potentially recommending they take over control. If a danger situation is perceived, the vessels would send out radio messages, sound signals and if necessary, take evasive action. Note that the operator's view from the remote stations would be as good as if they were onboard.

4.9 CERTIFICATION STANDARDS

Vessel and its command and control system would comply with, or exceed:

- Statutory - National (Swedish and / or Flag State) statutory requirements, including Collision Regs.
- Classification – recommended applicable standards and rules, and recommendations
- IMO – consideration given to any relevant recommendations
- Military – military standards if applicable or relevant
- School of Hard Knocks – input from experienced front-line SAR and marine personnel will be evaluated provided it is not in conflict with the foregoing.

4.10 IMPACT OF DESIGN FOR POLAR WATERS

The proposed design is significantly impacted by its proposed area of operation, ranging north from Newfoundland through the Labrador Sea up through Davis Strait into Baffin Bay between Baffin Island and Greenland. This area suffers from heavy weather, freezing temperatures leading to ice accretion, first year pack ice, slush ice (affecting water jets and engine cooling systems), multi-year ice hazards (such as growlers, bergy bits, ice bergs and multi-year ice in pack ice), limited repair and maintenance facilities, and so on. These factors affect the external vessel design, internal design, hull material and scantlings, machinery, and outfitting. Some other areas of the world have similar conditions, however, many do not. The vessel design could be tailored for the specific requirements, while maintaining the overall system needs and requirements.

4.11 FUEL CAPACITY AND RANGE

As previously noted, the operational profile could be:

- Proceed at moderate, fuel-efficient speed, from base port to its 'station' (alternatively may be towed);
- Loiter on station for required length of time, for example, 30 days;
- Return at moderate, fuel-efficient speed, from 'station' to base port (alternatively may be towed);
- If emergency response required, proceed at maximum speed (considering sea conditions) to the incident site, remain on site to render assistance as required, proceed at reasonable speed to port of refuge.

As an example, for a 'station' off northeastern Labrador, the distance from base port / port of refuge to station (moderate speed) could be 400 nm, radius of operation (high speed) up to 200 nm. Allowing for up to three days on an incident site, and 30 days loitering on station, and a reasonable reserve, fuel capacity could be calculated. Using wind, sea state and current data from Pilot Charts, daily fuel consumption to remain on station can be calculated, as well as expected operational response capabilities (due to weather impact).

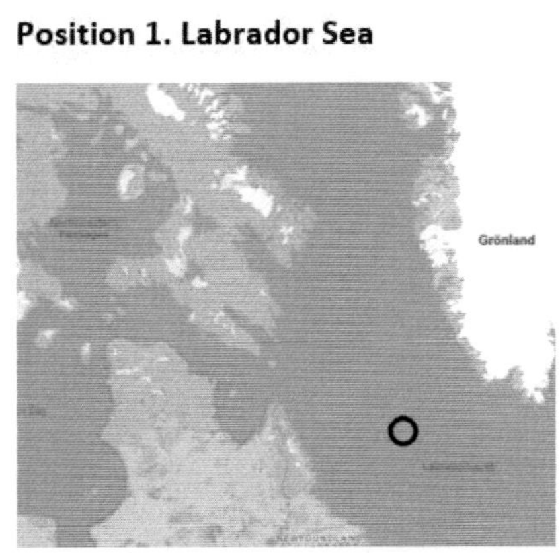

Figure 11: Proposed location of a LHSRV (Google Maps)

4.12 SUMMARY VESSEL CHARACTERISTICS

Proposed LHSRV characteristics are:

- Length 20m
- Breadth 5m
- Draft 2m
- Full load displacement 35 tons
- Maximum speed 40 knots

4.13 COST–BENEFIT ANALYSIS

Although a comprehensive Cost-Benefit Analysis is outside the scope of the current study, a few key points are noted here.

From the Benefit perspective, providing adequate and effective Search and Rescue coverage has many social and economic benefits, in terms of perceptions of safety, economic activities (such as fishing) being encouraged, pollution costs avoided, and lives saved.

From the Cost perspective, providing rescue coverage in Remote Regions using the Rescue Response System of Loitering Optionally Crewed High Speed Rescue Vessels shows significant savings in comparison to other offshore capable rescue vessels. Based on a preliminary evaluation, on top of the LHSRV Rescue System direct capital and operational costs, it would have to include an allowance for 'local' maintenance and related costs. However, the LHSRV would also not put rescue ships' crews lives at risk, and would result in lower emissions, which are additional benefits.

Admittedly, the LHSRV would not have all the capabilities of a fully crewed rescue ship, or the towing capabilities of an Emergency Response Tug, but as is the case of many marine areas, such as the Labrador Sea, the option may often be the Loitering Optionally Crewed High Speed Rescue Vessel or nothing. This is not necessarily an either / or situation, as the LHSRV could be part of a SAR Response System, working in conjunction with other larger vessels but offering a closer high speed first on scene capability.

> *In all cases in the continuum of rescue response, which ranges from nothing to being in the emergency room of a major medical facility, the uncrewed high speed rescue vessel would likely fall somewhere between nothing and a manned rescue boat. However, to a person in the water, the uncrewed high speed rescue vessel would most likely appear to be a massive improvement on nothing.* [2]

5. CONCLUSIONS

5.1 THE WAY FORWARD

The need, for Maritime Search and Rescue Response in a Remote Region, has been presented. A viable solution has also been presented, the Docksta Loitering Optionally Crewed High Speed Rescue Vessel System (consisting of one or more vessels). As part of a Search and Rescue system, these vessels could contribute to an efficient and effective capability. In a Coast Guard role, in addition to their Search and Rescue role, by their presence, through monitoring traffic they could also enhance national sovereignty and security in regional seas.

The concept is a Rescue Response System, with autonomy and remote controls while in the loitering state, and regional or local remote controls, or even onboard control, while in the rescue response state. The physical design of the vessels could differ from what has been proposed and could even vary within the same operation group, depending on the needs.

The person in the water recovery system may require further consideration. The system may vary from user to user, or even from vessel to vessel in one user group, depending on needs and user preferences. However, all systems would require seaworthiness, reliability, ease of maintenance and repair, and the confident support of regional SAR personnel.

The command and control system requires robustness, reliability, and freedom from hacking threats, while at the same time being effective in terms of emergency response, efficient (cost-effective) and reliable from both national security and helping to meet international rescue response obligations.

The authors explored this concept further with potential operating partners at the World Maritime Rescue Congress (Poster Presentation) and at the RINA Surv11, Surveillance, Search and Rescue Craft Conference, Rotterdam, June 18-22, 2023.

5.2 FINAL THOUGHTS - FROM CONCEPT TO REALITY: THINK SLOW, ACT FAST [1]

As noted in the preamble, successful implementation of a project, such as this, requires careful consideration by experienced, competent personnel at all levels:

- Definition of needs and requirements;
- Technical design;
- Regulatory compliance;
- Front-line operations.

In particular, the involvement of potential operators, Maritime Rescue Societies, Coast Guards and similar organizations, is required to:

- Refine the Concept of Operations, taking it from this Feasibility Study stage to an initial application stage;
- Determining areas of operation;
- Revising the design, if required, to best suit operational needs;

- Involving national SAR partners, such as Rescue Coordination Centres, SAR response (Coast Guards, Rescue Societies, Air SAR response), at both the planning and front-line response levels, other stakeholders such as ship operators, regional and local levels of government, and others who may have an interest, or be able to provide input;
- Confirming regulatory requirements;
- Establishing operating procedures and protocols.

After this, the next steps would be to:

- Establish contractual relationships with one or more SAR organizations and / or national authorities.
- Start small; put into operation one or two vessels, evaluate operations, and design, initial operational hurdles, revise as required.
- Full Scale Operation; put into operation, ongoing evaluation and problem solving – on a National or International level.

6. REFERENCES

1. FLYVBJERG, B., GARDNER, D., Chapter 1 'Think Slow, Act Fast', *How Big Things Get Done*, 2023.

2. Unmanned Vehicles could aid Search and Rescue, The Maritime Executive, Dec 16, 2016. https://maritime-executive.com/editorials/unmanned-vehicles-could-aid-search-and-rescue

3. DALZIEL, J., PELOT, R., The Application of Unmanned Marine Vehicles to Maritime Search and Rescue, SURV9 – *Surveillance, Search and Rescue Craft Conference, London,* April 18, 2018.
4. Robotic Search and Rescue: the new ICARUS (Integrated Components for Assisted Rescue and Unmanned Search Operations) system ready for use from 2016, NATO CMRE, July 15, 2015. https://www.cmre.nato.int/rockstories-blog-display/321-cmre-at-the-forefront-of-nato-s-interoperability-2

5. DALZIEL, J., Future Technology and Preserving Life at Sea, SURV10 – *Surveillance, Search and Rescue Craft Conference, London*, January 13, 2021.

6. DALZIEL, J., PELOT, R., SHERRARD, T., Remote Region Search & Response – Air Deployed Remotely Controlled Marine Craft, *RINA Autonomous Ships Conference, London* June 18, 2020.

7. SOLAS Chapter V, Regulation 7, IMO, retrieved from https://assets.publishing.service.gov.uk/government/uploads/system/uploads/attachment_data/file/343175/solas_v_on_safety_of_navigation.pdf

8. SAR Convention, IMO https://www.imo.org/en/OurWork/Safety/Pages/SARConvention.aspx

9. JAIN, V., SAINI, D., GUPTA, M., JOSHI, N., MISHRA, A., BANSAL, V., and HEMANTH, D.J. A Comprehensive Review on Design of Autonomous Robotic Boat for Rescue Applications, *Hindavi Mathematical Problems in Engineering, Volume 2021, Article ID 6614002*, June 22, 2021.

10. MATOS, A., SILVA, E., ALMEIDA, J., MARTINS, A., FERREIRA, H., FERREIRA, B., ALVES, J., DIAS, A., FIORAVANTI, S., BERTIN, D. and LOBO, V. Unmanned Maritime Systems for Search and Rescue, in Search and Rescue Robotics - From Theory to Practice, *INTECH*, 2017.
11. Owner of Atlantic Charger, sunk in Frobisher Bay, angry no helicopter responded, CBC, Oct 9, 2015 https://www.cbc.ca/news/canada/newfoundland-labrador/atlantic-charger-brad-watkins-skipper-rescue-1.3263361
12. 'We were damn near killed': Atlantic Charger captain speaks out on survival, rescue, CBC, Oct 14, 2015 https://www.cbc.ca/news/canada/newfoundland-labrador/atlantic-charger-captain-1.3269974#SnippetTab
13. Atlantic Charger lost, but rescued fishing crew coming home from Frobisher Bay, CBC, Sept 22, 2015 https://www.cbc.ca/news/canada/nova-scotia/atlantic-charger-fishermen-navy-rescue-1.3238219

14. TSB report into Saputi Incident, TSB Canada, Feb 16, 2016 Marine Transportation Safety Investigation Report M16C0016 - Transportation Safety Board of Canada (tsb.gc.ca)

15. Reporting after near loss of large shrimp trawler, Saputi. CBC, March 7, 2016. Nunavut officials press for Arctic search and rescue base | CBC News

16. SAAB website, Combat Boat CB90 Next Generation https://www.saab.com/products/docksta-cb90

17. Unmanned systems tested by the Swedish Navy during exercise, Sept 8, 2022 https://www.saab.com/newsroom/stories/2022/september/unmanned-systems-tested-by-the-swedish-navy-during-exercise

7. AUTHOR'S BIOGRAPHY

Emil Jonze, M.Sc., Eng. Phys, studied non-invasive measurement technologies in Sweden at Umeå University. He has worked with computer vision and control systems within variety of fields; such as automotive, autonomous system, farming and mining. Over the last two years he has worked within SAAB Kockums with USV/ASV systems, with the role as product leader for unmanned system.

John W. Dalziel, M.Sc., P.Eng., MRINA, studied Naval Architecture in the United States and Scotland, and Management Science in England. He has sailed through the Arctic with the Canadian Coast Guard, and with the fishing industry. Over the past half century, most of his experience has been in the supervision of ship construction, refits, and repairs, and in safety inspection and survey on behalf of Flag State and Class. He has spoken internationally on maritime safety, most recently at the World Maritime Rescue Congresses, Bremerhaven, Germany, 2015, and Vancouver, 2019, and at SURV9 and SURV10. As an Adjunct Professor in the Department of Industrial Engineering, Dalhousie University, Halifax, Canada, he is investigating 'The Application of Unmanned Marine Vehicles to Maritime Search and Rescue'.

Ronald Pelot, PhD., P.Eng., has been a Professor in the Department of Industrial Engineering at Dalhousie University since 1994. In 1997, he founded the Maritime Activity and Risk Investigation Network at Dalhousie (now rebranded as MARS). Since then his team has developed new software tools and analysis methods applied to maritime safety (accidents), coastal zone security, and marine spills. Research methods encompass spatial risk analysis, vessel traffic modelling, data processing and pattern analysis, location models for resource allocation, and safety analyses. Key projects include examining the effects of extreme weather on fishing safety, optimal resource allocation for Search & Rescue, and arctic shipping risks and response. Dr. Pelot has published over 50 journal articles and produced more than 100 technical reports.

FULL SCALE VALIDATION OF A FREQUENCY DOMAIN GYROSCOPE ROLL STABILISING MODEL ON VARIOUS FAST CREW SUPPLY VESSELS

G J Poen and **A A K Rijkens**, Damen RD&I BV, Netherlands

SUMMARY

Roll motions on board can have a big impact on human comfort and operational limits. Roll stabilising gyroscopes are effective systems to reduce roll, particularly at zero speed. Seakeeping simulations are often performed in the frequency domain to keep simulation times practicable. This paper presents a method of including a gyroscope as roll damping device, by adding a seventh degree of freedom to the rigid body motions of the hull, with particular emphasis on corrections for the non-linear behaviour of the gyroscope due to large angles and mechanical limitations. The method is finally validated against full scale sea trial measurements on a variety of fast crew supply vessels and gyroscopes. The results match very well for zero speed, but at high speed both measurements and calculations are less accurate and larger deviations are found between the two.

1. INTRODUCTION

High comfort levels on board are part of the requirements for any new vessel. A big part of comfort on board is related to motions and seasickness. To quantify the comfort level onboard, requirements are typically set for accelerations in various directions, motion sickness incidence (MSI) and motion induced interruptions (MII) and apply to both crew and passenger accommodations and workstations in various degrees. Of all six hull rigid body motions, it is roll that stands out as the one that can be severe at both zero speed and underway. The issue with rolling of ships, particularly monohulls, is the lack of inherent roll damping. This can cause large roll angles in beam or quartering waves, especially when wave periods are close to the natural roll period of the vessel. Large roll amplitudes impact transverse accelerations and MII levels mainly on higher deck levels, e.g. in the wheelhouse, especially on ships with a high GM and low natural roll period. MSI levels are determined by vertical and transverse accelerations, though the influence of transverse accelerations is not yet widely implemented in MSI formulas. Nevertheless, adding a roll reduction device to a monohull vessel can improve comfort onboard and is therefore quite common and often necessary to comply with these criteria.

The various types of roll reduction devices that have been developed range from simple passive measures, e.g. bilge keels, to actively controlled devices such as gyroscopes, stabilizer fins, magnus rotors and trim flaps or interceptors. Of the active devices it is the gyroscope that performs best at zero speed, while still performing well at forward speed.

The performance of the gyro at zero speed makes it particularly suitable for fast crew boats that spend some time at an offshore location at zero speed while transferring people. Fast crew boats are typically lightweight to minimize resistance, so it is important to select a gyro of proper size and weight, not to mention cost. In the design phase of a new ship, or even in a retrofit, it is therefore useful to be able to run seakeeping simulations for multiple combinations of ship loading condition, gyro type and environmental conditions, e.g. wave height, period and heading. Time domain simulations can be very accurate, but they take a lot of computational effort. Instead, the linear frequency domain is much more suited for large amounts of simulations. The one drawback is that linear calculations tend to go awry in higher waves due to the lack of implementing upper limits of roll reduction devices. Therefore, a frequency domain method is proposed for gyros, which includes correction factors for non-linear behaviour and mechanical limitations. The method is validated with the use of full scale sea trial data.

2. GYROSCOPE FUNDAMENTALS

Gyroscopes suitable for roll reduction of a ship typically weigh approximately 3% of the ship's displacement. The largest gyroscopes in production can deliver moments up to 1 MNm and weigh in excess of 20 ton, see Figure 1 (left) for an example. A gyroscope is in essence a spinning mass—inside a near vacuum to reduce air friction—which can rotate around an axis perpendicular to the spin axis. That rotation is called precession (denoted by α) and the precession (angular) velocity $\dot{\alpha}$ is what generates gyro roll stabilisation. The way it works is described below, see also Figure 1 (right) for a schematic of the terms.

- On board a flywheel is spinning at spin rate ω_{spin} and with mass moment of inertia J_{spin}.

- The angular momentum of the gyro is $K = J_{spin} \times \omega_{spin}$.
- The ship is subjected to waves and rolls with a certain roll angular velocity around the x-axis, $\dot{\Phi}$.
- Due to the gyroscopic effect, a precession moment is generated on the gyro perpendicular to both spin-axis and roll-axis equal to $M_{prec} = K \times \dot{\Phi}$.
- The gyro precession behaves as a mass-spring-damper with a precession mass moment of inertia (mostly the spinning mass and its surrounding shell), precession damping (friction in the precession bearings) and precession stiffness (most gyros are made with their vertical centre of gravity just below the precession axis).
- The precession moment M_{prec} therefore results in a precession velocity $\dot{\alpha}$.
- Due to the gyroscopic effect, a gyro moment is generated on the gyro perpendicular to both spin-axis and precession-axis equal to: $M_{gyro} = K \times \dot{\alpha}$. This moment works in the opposite direction of the initial roll velocity. It therefore reduces the roll.
- The gyro moment acts purely on roll when the precession axis is vertical, i.e. when $\alpha = 0$. This coincides with the region of precession where the precession velocity is highest and therefore the gyro moment is highest. When the precession angle is large, the gyro moment largely acts in yaw direction and only partially in roll. However, at large precession angles the precession velocity approaches zero, which also brings the gyro moment to zero. Due to the low gyro moment and large yaw inertia, this does not impact course stability.
- To prevent the gyro precession from overshooting and hitting its mechanical limits, a precession control moment M_{cont} is added. The control moment is also tuned to the vessel and is governed by many parameters in reality.
- Pausing the gyroscope roll damping can be done by hydraulically locking the precession axis in any position. With $\dot{\alpha} = 0$ all gyro moments reduce to zero.

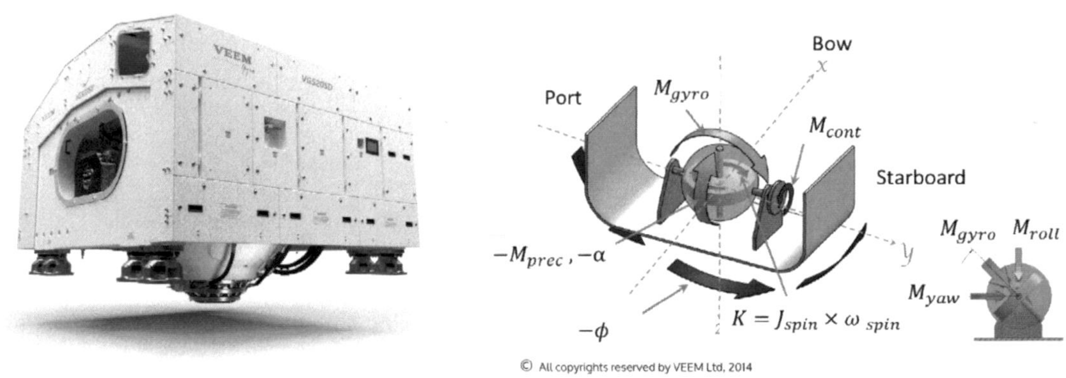

Figure 1: Left: VEEM VG520SD Gyro. Right: Gyro motion and moment definitions.

3. LINEAR EQUATION OF MOTION

The benefit of working in the linear frequency domain is that calculations are much faster than time domain simulations. The gyro can be added to the equation of motion of the rigid hull and the results can be obtained through spectral properties.

3.1 FREQUENCY DOMAIN RAO

To calculate the roll response of a ship in the frequency domain—with or without gyro—one must first calculate the response amplitude operators (RAOs). These are obtained by solving the equation of motion for regular waves of a range of frequencies. For a ship without gyro, there are 6 degrees of freedom (DOF); surge, sway, heave, roll, pitch and yaw. When adding a gyro there is 7th DOF: precession. The equation of motion can then be written as in Equation (1).

$$\left(-\omega^2[\mathbf{M}+\mathbf{A}(\omega)]+i\omega[\mathbf{B}(\omega)]+[C]\right)\begin{bmatrix}\hat{X}\\ \hat{Y}\\ \hat{Z}\\ \hat{\Phi}\\ \hat{\Theta}\\ \hat{\Psi}\\ \hat{\alpha}\end{bmatrix}=\begin{bmatrix}\hat{F}_1\\ \hat{F}_2\\ \hat{F}_3\\ \hat{M}_4\\ \hat{M}_5\\ \hat{M}_6\\ \hat{M}_{cont}\end{bmatrix} \tag{1}$$

Strictly speaking the motions in the XZ-plane (surge, heave and pitch) are uncoupled from roll and could be disregarded. The only coupling with the gyro happens between roll and precession and is shown in Equation (2).

$$\left(-\omega^2\begin{bmatrix}I_{xx}+a_{44}(\omega) & 0\\ 0 & I_g\end{bmatrix}+i\omega\begin{bmatrix}b_{44}(\omega) & K\cos(\alpha)\\ K\cos(\alpha) & b_g\end{bmatrix}+\begin{bmatrix}c_{44} & 0\\ 0 & c_g\sin(\alpha)/\alpha\end{bmatrix}\right)\begin{bmatrix}\hat{\Phi}\\ \hat{\alpha}\end{bmatrix}=\begin{bmatrix}\hat{M}_4(\omega)\\ \hat{M}_{cont}(\omega)\end{bmatrix} \tag{2}$$

Note that this contains two cosine terms and one sine term. Traditional linearisation will simplify the cosine term to 1 and the $\sin(\alpha)$ to α. However, for angles larger than approximately 20 degrees this is no longer valid. To be able to operate linearly even for precession angles up to 70 degrees, each of these terms is replaced with a constant correction factor. Further attention to this is given in Section 4.2. The number of active gyros can be taken into account through the integer n_g. Equation (2) then simplifies to Equation (3).

$$\left(-\omega^2\begin{bmatrix}I_{xx}+a_{44}(\omega) & 0\\ 0 & I_g\end{bmatrix}+i\omega\begin{bmatrix}b_{44}(\omega) & n_g K\beta_{\cos}\\ K\beta_{\cos} & b_g\end{bmatrix}+\begin{bmatrix}c_{44} & 0\\ 0 & c_g\beta_{\sin}\end{bmatrix}\right)\begin{bmatrix}\hat{\Phi}\\ \hat{\alpha}\end{bmatrix}=\begin{bmatrix}\hat{M}_4(\omega)\\ \hat{M}_{cont}(\omega)\end{bmatrix} \tag{3}$$

The hull (added) mass, damping, stiffness and wave excitation forces are calculated with Shipmo, a strip theory solver. The gyro precession damping b_g is set to critical damping, i.e. $b_g=\sqrt{4I_g c_g}$. The system is stable for $b_g>0$ and non-oscillating in free precession for $b_g\geq\sqrt{4I_g c_g}$, see also Perez and Steinmann [1].

3.2 RESPONSE SPECTRUM AND RMS

The RAO is combined with the wave spectrum to form the response spectrum. The wave energy density spectrum $S_\zeta(\omega)$ is defined by significant wave height H_s, peak period T_p, peakedness factor γ. The wave spectrum can be spread out over several headings θ using cosine spreading, creating a two-dimensional wave spectrum $S_\zeta(\omega,\theta)$. The response spectrum for roll is then obtained according to Equation (4). The property to evaluate roll and roll reduction is the Root Mean Squared (RMS) value of the roll response. The RMS value can be derived from a (response) spectrum using Equations (5) and (6). Precession and any other response can be calculated in a similar way.

$$S_\Phi(\omega,\theta)=\left|\frac{\Phi_a}{\zeta_a}(\omega,\theta)\right|^2\cdot S_\zeta(\omega,\theta) \tag{4}$$

$$m_{0\Phi}=\iint S_\Phi(\omega,\theta)\,d\omega\, d\theta \tag{5}$$

$$RMS\ \Phi=\sqrt{m_{0\Phi}} \tag{6}$$

3.3 ROLL LINEARISATION

The hull roll damping $b_{44}(\omega)$, calculated by strip theory, is in reality strongly nonlinear with roll amplitude due to the large contribution of viscous damping. A way to deal with this, is to calculate the roll results for several values of $b_{44}(\omega)$, each belonging to a different roll amplitude, see Figure 2. The best practical value of $b_{44}(\omega)$ is the value that corresponds to twice the RMS of the roll signal, which can be found by intersecting the 2*RMS roll results and $b_{44}(\omega)$ amplitude curves.

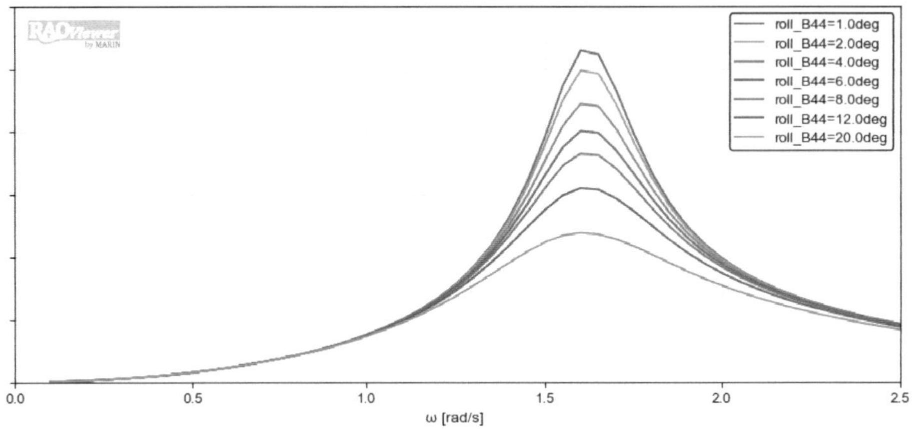

Figure 2: Roll RAO for roll damping values belonging to several roll amplitudes.

4. NON-LINEAR CORRECTIONS

One of the biggest downsides of working in the frequency domain, is that the system will behave linearly no matter how large the input. In reality, a gyro will never precess more than its mechanical limit α_{limit} will allow (typically around 70 degrees). In a linear model the precession may reach hundreds of degrees for a ship in high waves, without any effect of these large—and impossible—angles. Although Perez and Steinmann [1] mentioned this issue, solutions were not part of the scope of their paper. In this study several corrections are implemented in the method to account for these non-linear effects.

4.1 RMS OF A TRUNCATED PRECESSION SPECTRUM

The simplest and most effective way to judge the precession behaviour in the frequency domain is by looking at the RMS. When calculating the precession RMS without any gyro control, the value might be anywhere from $RMS_\alpha \ll \alpha_{limit}$ to $RMS_\alpha \gg \alpha_{limit}$. For $RMS_\alpha \ll \alpha_{limit}$, a Rayleigh distribution of peaks will have a negligible amount of peaks exceeding α_{limit}, i.e. the distribution is not affected by the precession limit. The distribution of α can therefore be accurately modelled as a normal distribution. This is demonstrated by comparing full scale measured precession data of a gyro in low waves, with an RMS_α of only 12.6 degrees, to a normal distribution with the same RMS, see Figure 3 (left). When waves are relatively high, the RMS value will get closer to the limit value. That means in reality a significant amount of peaks will be reduced to avoid exceeding the mechanical limit. In the extreme case, the precession motion approaches a regular sine wave with precession amplitude equal to the limit value. This is demonstrated in Figure 3 (right), where the measured precession angles for high waves have an RMS of 43.4 degrees and are compared to a normal distribution and sine wave of equal RMS (note that the amplitude of the sine wave equals $43.4 \cdot \sqrt{2} = 61.4\, deg$). The measurement shows much more similarity with the sine wave than with the normal distribution.

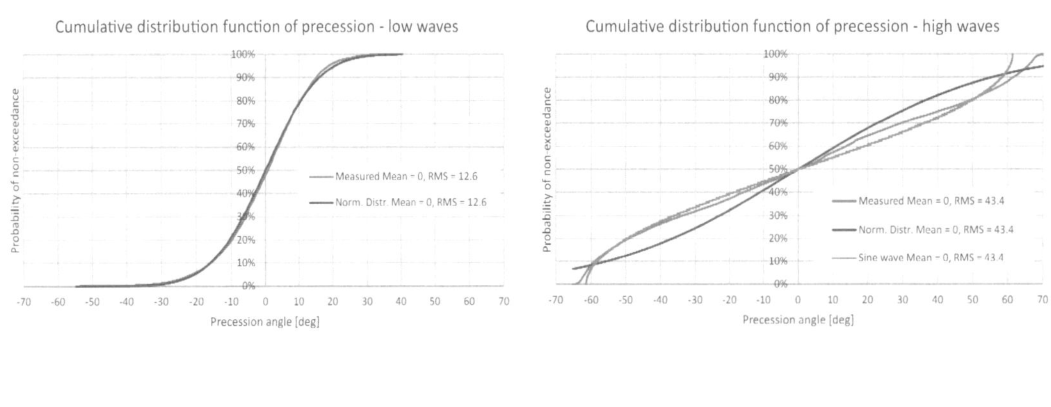

Figure 3: Precession angle distribution in low (left) and high (right) waves.

The way to incorporate this fact in the frequency domain calculations is done by setting a target RMS precession value and iterating to reach that target. Depending on the ratio of unrestricted precession RMS to precession limit, the precession limit will truncate anything from close to none to close to all peaks in the spectrum. This is illustrated in Figure 4. By numerically dividing the truncated spectrum in n bins of amplitude X and probability of occurrence Pr (X), the truncated RMS precession, i.e. the target RMS value, can be calculated according to Equation (7). Note that for the extreme case, i.e. all bins having amplitude $X = \alpha_{limit}$, the formula simplifies to the RMS of a sine wave according to $RMS_{target} = \sqrt{2}/2 \cdot \alpha_{limit}$.

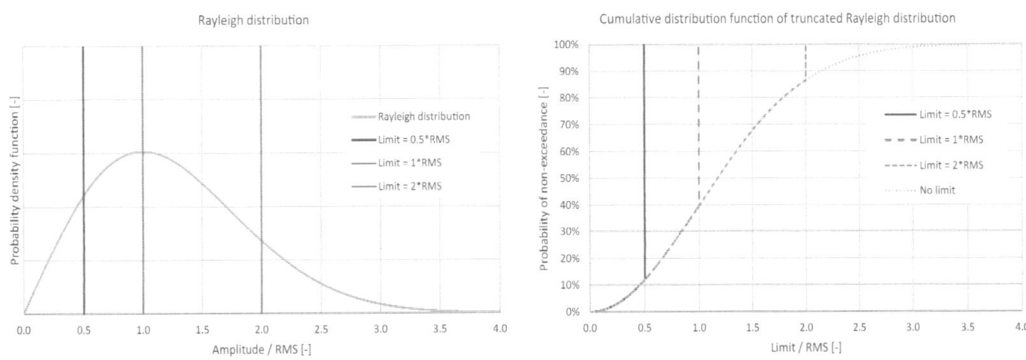

Figure 4: Precession Rayleigh distribution including truncation at various relative limit values.

$$RMS_{target} = \frac{\sqrt{2}}{2} \cdot \sqrt{\sum_{i=1}^{n} \Pr\big(X(i)\big) \cdot X(i)^2} \qquad \textbf{(7)}$$

This limitation is only applied on precession angle, but not on precession velocity which is linearly proportional to the gyro moment. All gyros have a maximum gyro moment, but this is typically achieved at fairly short roll periods and does not exceed the restrictions on precession angles by much, if it all. For gyros with a relatively large angular moments and low gyro moment limit, this might become important though.

4.2 LARGE PRECESSION ANGLES

Large precession angles, even when having realistic RMS values, are still outside the linearisation range of approximately 20 degrees. To correct for these large angles in the sine and cosine terms, the difference between non-linear sine and cosine terms from Equation (2), i.e. $\sin(\alpha)$ and $\dot{\alpha}\cos(\alpha)$, and linear sine and cosine terms, i.e. $\sin(\alpha) = \alpha$ and $\dot{\alpha}\cos(\alpha) = \dot{\alpha}$, was investigated for several full scale measured precession signals with RMS values ranging from small (12 deg) to large (50 deg). The difference in resulting RMS of both non-linear and linear approach yielded correction factors β_{sine} and β_{cosine}, see Figure 5. By looking up the RMS_{target} from Equation (7), the proper correction factor can be found and inserted in Equation (3).

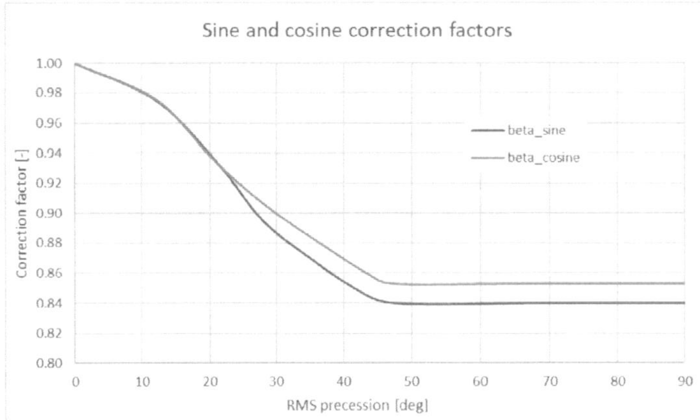

Figure 5: Sine and cosine correction factors for large precession angles.

4.3 GYRO CONTROL MOMENT

After obtaining a target RMS value for precession and finding the corresponding sine and cosine correction factors, the precession control Moment M_{cont} needs to be adjusted to achieve the target RMS. Perez and Steinmann [1] propose a control moment depending only on precession information in the form of Equation (8).

$$M_{cont} = b_{cont} \cdot \dot{\alpha} + c_{cont} \cdot \alpha \tag{8}$$

Inserting Equation (8) into Equation (3) leads to Equation (9). By iteratively increasing b_{cont} and c_{cont} the RMS precession will drop to the target RMS. The ratio between b_{cont} and c_{cont} can be tuned to match the full scale results in the best way. Note that $b_{cont} \geq 0$ and $c_{cont} \geq 0$ to maintain stability.

$$\left(-\omega^2 \begin{bmatrix} I_{xx} + a_{44}(\omega) & 0 \\ 0 & I_g \end{bmatrix} + i\omega \begin{bmatrix} b_{44}(\omega) & n_g K \beta_{\cos} \\ K\beta_{\cos} & b_g + b_{cont} \end{bmatrix} + \begin{bmatrix} c_{44} & 0 \\ 0 & c_g \beta_{\sin} + c_{cont} \end{bmatrix}\right) \begin{bmatrix} \hat{\Phi} \\ \hat{\alpha} \end{bmatrix} = \begin{bmatrix} \hat{M}_4(\omega) \\ 0 \end{bmatrix} \tag{9}$$

5. SEA TRIAL VALIDATION

Several trials were held with gyroscopes on a range vessels. The measured data from these trials is compared to the results of the calculations under the same conditions.

5.1 VESSELS AND GYROSCOPES

Sea trials were done on a range of Fast Crew Supply (FCS) vessels, see Figure 6, in combination with several gyros, see Table 1 for an overview. Dedicated gyro trials were held on an FCS 1905 and FCS 5009, where the gyros were placed on deck temporarily. Note that it does not matter where one places a gyro on board for the roll damping performance of the gyro. Permanent VG520SD gyros were fitted on an FCS 5009 and FCS 7011. All trials were held in the North Sea, except for the FCS 5009 with VG520SD, which was held in the Gulf of Mexico.

Figure 6: Full scale trials. From left to right: FCS 1905, FCS 5009, FCS 7011.

Table 1. Overview of sea trials.

Ship type	Displacement	Gyro type	Angular momentum
[-]	[ton]	[-]	[N.m.s]
FCS 1905	49	1 xSeakeeper 21000	1 x 21 000
FCS 1905	49	2 x Seakeeper 21000	2 x 21 000
FCS 5009	405	2 x VG100SD	2 x 100 000
FCS 5009	462	1 x VG520SD	1 x 520 000
FCS 7011	496	1 x VG520SD	1 x 520 000

5.2 SEA TRIAL DATA

The trials were each performed in varying wave conditions and speeds. The tests all consisted of two measurements under the same circumstances; a roll measurement without gyro and a roll measurement with gyro. The roll was measured with an Xsens MRU, logged at 10 Hz. The precession motions of the gyros were also logged—mostly using internal sensors of the gyro—at 10Hz or more. The wave conditions were obtained from Rijkswaterstaat wave buoy data in the North Sea and Copernicus Marine in the Gulf of Mexico.

The loading condition of the ship was calculated from light ship weight and deadweight inspection. The GM of the vessel was calculated in Shipmo and based on the linesplan and centre of gravity of the ship. The roll damping of the hull (without gyro) was tuned so that the calculated roll without gyro matched the measured roll without gyro.

A number of difficulties arose during the sea trials, that are all related to the environmental input.

- Uncertainty in wave height. Waves are ideally in a steady state, but often the waves are still either growing or dying out. Even according to wave buoy measurements, the wave height on the North Sea sometimes showed a 20% increase in a matter of 20 minutes.
- Uncertainty in wave periods. Similar to wave heights, the period varies. Especially when close to resonant roll, the impact can be very large.
- Uncertainty in wave heading. In reality there will always be some degree of wave spreading and in some cases more than a little. It's very hard to estimate the degree of spreading, though.
- Uncertainties associated with speed. The main problem with measuring at speed is the distance covered during the measurement; the waves may differ between start and end. Examples of this happened mostly on the North Sea, where there was a rather short stretch available between shore and restricted wind parks. The waves were noticeably lower close to shore. In one trial—that was subsequently not taken into account in the validation—the waves even noticeably changed from beam seas to bow quartering during a single measurement. It's also not possible to combine runs on the return heading, as waves will never be perfectly oriented beam-on and the encounter frequency will noticeably shift between the two runs. Going back to redo the same stretch in the same direction takes time which introduces more risk on wave variations.
- Wind gusts. Wind is not taken into account in the equation of motion at all. The wind not only creates a mean roll (heeling) angle, which had to be corrected for in all trials, but the gusts were sometimes affecting the roll angles strongly at low frequency intervals. This can lead to different environmental conditions between gyro-off and gyro-on measurements. Even if the wind is equally gusty during both gyro-off and gyro-on measurements, the gyro-off roll response is initially underpredicted in the simulations (based on waves only) and as a consequence the hull roll damping is lowered. The sensitivity to wind gusts varied between trials and ship types.

Given the relatively large uncertainty on environmental conditions, it is hard to find an acceptable amount of inherent hull roll damping per ship type that matches all gyro-off measurements. Some trial results were disregarded as they proved useless—some even showed significant increase in roll with gyro on. The problem with miscalculating the hull roll damping, is that the roll reduction performance of the gyro is affected; adding a roll damping device to an already well dampened system will do much less than adding one to a very lightly dampened system.

5.3 RESULTS

There are a few parameters that can be tuned in the calculation. These are principally the maximum precession angle and the ratio between b_{cont} and c_{cont} in Equation (9). The maximum precession angle is given in specifications, but in reality all gyros are protected against hitting their mechanical limits so it fits the results better to reduce the maximum precession angle a little in the calculations. The results show a better fit to the measurements with a strong emphasis on b_{cont}.

The calculations show a realistic trend with varying wave height and wave period. An example is shown in Figure 7, which contains a wide range of Hs-Tp variations for the FCS 5009 with 1 x VG520SD, as well as the measured result during sea trials. For higher waves the gyro performance starts to degrade, thanks for the peak reduction correction factor. The maximum roll damping occurs around the natural roll period of the vessel of approximately 5.2 seconds, as response is dominated by damping. For long waves the roll period becomes equally long and the precession velocity and gyro moment go down. In case of long waves the wave slope and ship stiffness become dominant for roll response. For short waves the inertia of the ship will dominate the roll response.

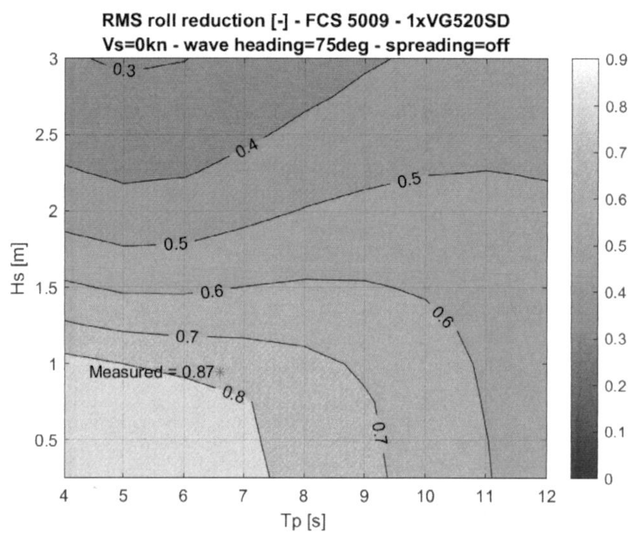

Figure 7: FCS 5009 RMS roll reduction calculated, with measurement result shown.

The full set of trial results with measurement values and calculation results are shown in Table 2, which gives the conditions as well as the measured RMS roll reduction (R.R. meas.) and calculated RMS roll reduction (R.R. calc.). A few trends are visible.

- The zero speed cases are all quite accurate, with less than 15 percent point difference. These are shaded in light green in the table. The one exception is the FCS 7011 result with a difference of 29 percent point. The most probable cause is the strong wind gusts (average wind speeds were 28kn but gusts were up to 50kn), which led to low hull damping in Shipmo to match the gyro-off results as well as larger differences in the conditions between gyro-off and gyro-on measurements. Note that the VG520SD was not yet properly tuned during these trials, leading to a reduced precession motion and reduced roll reduction performance.
- At forward speed the measured roll reduction drops significantly in almost all cases, whereas the calculated roll reduction stays roughly the same. This most notably the case for the FCS 1905 results at 20 kn. Although forward speed measurements have a larger uncertainty, see section 5.1, it seems that there is more at play here. They gyro fundamentals are not impacted by forward speed, so the most likely explanation is that there is a much larger difference between the calculated coefficients of the equation of motion under forward speed than at zero speed. Non-linear effects start to play a significant role for hull damping at higher speeds and frequency domain tools are not the best at handling this.
- The calculated roll reduction is almost always slightly higher than the measured roll reduction. This is probably due to some precession control parameters that are not modelled at all as well as aforementioned differences at higher speed.
- Note that stern quartering waves (relative wave headings of <90 deg or >270 deg) at speed pose the additional challenge of zero-encounter frequencies, where for some frequencies the wave speed is equal or close to the ship speed. Frequency domain solvers such as Shipmo typically wildly

overpredict the response in the horizontal plane (surge, sway, yaw), due to lack of any restoring force, i.e. stiffness term. Through coupling terms the roll motions suffer as well for these frequencies.

Table 2. Sea trial measured and calculated results. Note: FCS 7011 trials were done with a sub-optimal tuned gyro.

Ship type	Gyro type	Heading	Vs	Hs	Tp	R.R. meas.	R.R. calc.
[-]	[-]	[deg]	[kn]	[m]	[s]	[%]	[%]
FCS 1905	1 x Seakeeper21000	90	0	1.00	4.8	67%	75%
FCS 1905	1 x Seakeeper21000	90	8	1.10	5.3	46%	73%
FCS 1905	1 x Seakeeper21000	90	0	1.20	7.7	60%	75%
FCS 1905	1 x Seakeeper21000	90	8	1.30	7.1	37%	71%
FCS 1905	2 x Seakeeper21000	90	0	1.00	4.8	84%	90%
FCS 1905	2 x Seakeeper21000	90	8	1.10	5.3	74%	90%
FCS 1905	2 x Seakeeper21000	90	20	1.20	5.0	20%	89%
FCS 1905	2 x Seakeeper21000	90	0	1.20	7.7	77%	89%
FCS 1905	2 x Seakeeper21000	90	20	1.20	7.1	24%	88%
FCS 1905	2 x Seakeeper21000	90	8	1.30	7.1	57%	88%
FCS 1905	2 x Seakeeper21000	135	9.5	1.50	7.7	68%	85%
FCS 5009	2 x VG100SD	90	0	1.00	5.7	33%	46%
FCS 5009	2 x VG100SD	90	8	1.10	5.7	35%	44%
FCS 5009	2 x VG100SD	90	23	1.20	5.7	29%	43%
FCS 5009	1 x VG520SD	75	0	0.95	6.6	87%	80%
FCS 5009	1 x VG520SD	105	19	0.90	6.3	60%	82%
FCS 7011	1 x VG520SD	90	0	1.60	5.7	50%	79%
FCS 7011	1 x VG520SD	150	19	1.60	5.8	32%	30%
FCS 7011	1 x VG520SD	60	19	1.60	5.9	0%	13%
FCS 7011	1 x VG520SD	115	19	1.70	6.3	25%	21%

6. CONCLUSIONS

In this study a frequency domain method of including gyroscope roll damping is proposed and validated against sea trial measurements. The method is an extension on the 6 degree of freedom rigid hull motions with a 7th for gyro precession. The benefit of working in the frequency domain is the high calculation speed, which makes it possible to assess comfort criteria in many combinations of conditions. Particular attention was paid to including non-linear limitations of the gyro, to prevent an overly optimistic performance in higher waves. This was done by controlling the RMS values of the precession in a theoretical way, that was supported by full scale precession measurements.

The validation with sea trial results was successful for zero speed cases. The gyro performance is only slightly overestimated. The differences at high forward speed are much larger and the gyro performance is overestimated a lot.

One of the biggest challenges is to get to good measurements during sea trials, particularly in windy conditions with developing seas such as on the North Sea. It is expected that much more accurate measurements can be made in swells as opposed to seas. The other issue is the reliability at higher speeds, which suffers from more difficulties in both accurate measurements and accurate calculations in the frequency domain.

Altogether the tool proves to be a good way to include gyroscope roll damping in seakeeping calculations at zero speed. Gyroscopes themselves are able to reduce roll in moderate conditions up to 90%, making them highly effective in improving comfort and safety on board, as well as extending operational limits in waves.

7. ACKNOWLEDGEMENTS

The authors would like to express their gratitude to VEEM Marine for their theoretical and practical support.

8. REFERENCES

1. PEREZ, T., and STEINMANN, P D., ‘Analysis of Ship Roll Gyrostabiliser Control’, *Proceedings of the 8th IFAC*, 2009.

9. AUTHORS BIOGRAPHY

G J Poen holds the current position of Development Engineer at Damen Research, Development & Innovation B.V.. The Author is responsible for predictions on a range of topics, including resistance, speed, sound levels, vibration levels, ship motions and fatigue, as well as carrying out sea trials to gather full scale data on performance.

A A K Rijkens holds the current position of Program Manager at Damen Research, Development & Innovation B.V.. The Author is responsible for research to new ship systems and concepts in close collaboration with the Delft University of Technology.

RINA SURV11 – INVESTIGATION OF HYDROGEN AS PART OF A DUAL-FUEL APPROACH FOR CREW BOATS

J Calver, C Rident and **T Beard**, BMT, UK

SUMMARY

This paper evaluates hydrogen propulsion and storage solutions currently available in the context of application to crew boats and explores the feasibility and design impacts of implementation. This concludes that the most suitable hydrogen fuel installation for existing crew boat designs is currently an H2 dual-fuel high speed engine, supplied by containerised on-deck compressed H2 storage. This integrates well into existing engine rooms and offers the possibility to convert at refit. A high-level concept design is outlined utilising this arrangement, and its performance in terms of diesel consumption and speed is assessed and compared to existing baseline designs. The design shows significant savings (up to 60%) in diesel consumption against proven diesel-mechanical crew boat designs.

1. INTRODUCTION

In recent decades greenhouse and toxic gas emissions restrictions have been put in place by increasingly restrictive regulations enforced by governments and supranational organizations. Road transportation is on a clear path to complete decarbonisation, and the marine shipping industry is certain to follow.

The use of hydrogen in the marine industry is growing and there are some commercial applications for passenger and vehicle ferry propulsion on limited routes. Notable examples are the Hyseas project on its third development stage for a hydrogen propulsion ferry, the 75 pax hydrogen ferry "*Sea Change*" recently delivered to San Francisco Bay and the "*MF Hydra*", an 82m liquid hydrogen ferry entering service in Ryfylke, Norway in April 2023.

Challenges associated with low volumetric density and power generation technology presently make hydrogen difficult to use as the sole power source for smaller commercial vessels such as offshore crew boats, where the requirements for a high vessel speed combined with a significant operating range and endurance capability are best achieved through the use of engines and fuels with a high power and energy density.

Dual-fuel engines provide the flexibility to run using hydrogen and diesel as alternative or complementary fuel sources. This leads to significant reductions in emissions when compared to conventional diesel engines, whilst also reducing required hydrogen tank capacities onboard. Additionally, dual-fuel engines allow operation with reduced emissions inside emission control areas, while leveraging diesel fuel's higher energy density for operation outside of emission control areas.

This paper evaluates hydrogen propulsion and storage and storage solutions currently available in the context of application to crew boats and explores the feasibility and design impacts of implementation in a crew boat. This concept is then evaluated against existing conventional diesel-mechanical vessels.

2. CREW BOAT TYPICAL DESIGN/REQUIREMENTS

A fast crew boat vessel serves to transport crew and cargo to and from offshore installations. (See Figure 1 for a typical vessel). They are used primarily to support offshore oil and gas production units, an industry that is under extreme exposure to government and public attention on carbon emissions throughout its production and delivery activities. A green propulsion system for a crew boat would serve to meet the regulatory framework on emissions and to provide some public goodwill towards an essential industry such as offshore oil production.

Figure 1: Alkahfi Comfort - 42m Fast Crew Boat, designed by BMT and built by Penguin International

Typical Fast Crew Boat designs range in length from around 30 metres to 45m, with 42m being a widely built length. Almost exclusively full-aluminium monohull construction, maximum speed typically approaches 30 knots. The emphasis of the design is on efficiency, deadweight capacity and comfort, with many vessels featuring 'business-class' style seating saloons for between 50 and 100 offshore technicians. FCBs have a substantial deadweight carrying capacity for their size, in the order of 150 tonnes (for a 42m vessel).

Traditionally, Fast Crew Boats have been designed to compete with the short transfer times offered by helicopter personnel transfer, which has driven the vessel's maximum speed and therefore installed power. Recent increases in oil prices have reduced the economic viability of helicopter transfers, reducing the speed pressure on crew boat designs. This same increase in oil price is leading to calls for more efficient and economical vessels.

3. H2 TECHNOLOGY REVIEW

3.1 H2 POWER PLANTS

3.1 (a) Fuel Cells

Fuel cells produce electricity and heat through electrochemical reactions with hydrogen and air, instead of combustion to produce electricity as opposed to kinetic energy. They consist of two electrodes; a negative (anode) and a positive (cathode), sandwiched around an electrolyte. Depending on the electrolyte used, there are several types of hydrogen fuel cells available to choose from (Table 1), however not all types are suitable for marine applications. The most researched and developed for use on board ships are the Proton Exchange Membrane (PEM) fuel cells. PEM fuel cells however require a high level of hydrogen purity (99.999%), and as such only 'pure' H2 can be used.

Table 1. Types of fuel cells

	Electrolyte mobile ion	Operating temperature
PEM	H^+	50 – 100°C
Alkaline	OH^-	50 – 200°C
Phosphoric Acid	H^+	ca. 220°C
Molten Carbonate	CO_2^{3-}	ca. 650°C
Solid Oxide	O^{2-}	500 – 1000°C

Several manufacturers producing PEM fuel cells for marine applications have emerged on the market recently. The majority offers commercial solutions in the range of 200 – 325 kW. Table 2 compares the main characteristics of the most suitable candidates for this case study.

TECO 2030 FCM 400 has the highest gravimetric power density among the three options, whereas Ballard FCWave is the most compact module.

Table 2. PEM fuel cells for marine applications examples

	Ballard FCWave	**PowerCell Marine System 200**	**TECO 2030 FCM 400**
Power output (kW)	200	200	325
Dimensions (L x W x H) (m)	1.21 x 0.74 x 2.20	0.73 x 0.90 x 2.20	1.38 x 0.80 x 2.13
Dry weight (t)	1.00	1.07	1.39
Gravimetric power density (kW/t)	200	187	234
Volumetric power density (kW/m³)	0.010	0.007	0.007

Using fuel cells offers an element of future-proofing, as the installed systems can be upgraded to slowly phase out the reliance on diesel as fuel cell and battery technology develop. However, this comes at the significant expense of increased weight, and increased system complexity. Fuel cells are also limited in their maximum power, with a limited power rating and reduced efficiency as the power output nears the top boundary of the operational range. The response time of fuel cells to increase/decrease power output is in the order of minutes. Consequently, they are currently typically combined in a hybrid system with batteries, which provide the transient load response capability and recharge from the FC when suitable. Used in this manner, they can provide a similar manner of load response as diesel-hybrid systems.

3.1 (b) Hydrogen-fuelled Internal Combustion Engines (H2-ICE)

The use of hydrogen as an alternative fuel to power internal combustion engines provides weight and space savings by replacing the fuel cells and electric propulsion equipment with an internal combustion engine. It provides increased reliability by eliminating a component of the propulsive train, however, introduces a new issue whereby the system must be shut down if an H2 leak occurs.

There is little technical difference between well-established gas-fuelled engines and hydrogen-fuelled Internal Combustion engines (H2-ICE). Indeed, Rolls-Royce has demonstrated that a modified 4000-series natural gas engine can be successfully run on hydrogen. The primary challenge to overcome is in precisely controlling the ignition point of the fuel, and also the higher flame temperature, which requires some level of modification to withstand. This reliance on an existing mature technology is a key benefit, with production infrastructure already in place to deliver units. Despite this, H2-ICE engines are not yet widely available for marine applications; MAN states that they expect H2-ICE to be available by 2030.

While the relatively long response times of fuel cells are suited to providing a background load, with batteries typically used to provide transient load response, H2-ICE can be used as a standalone power source, capable of handling transient load demand without requiring batteries, significantly reducing total installed weight and spatial requirements.

The efficiency of using H2 in an ICE is around 20-25%. This is significantly lower than the efficiency of hydrogen Fuel Cells, which approaches 60%. However, H2-ICE can burn lower-grade hydrogen than fuel cells; as hydrogen production ramps up, compatibility with onshore production methods that produce impure fuel may be an advantage.

Although H2-ICE have an inherently lower maximum efficiency than Fuel Cells the mature basis of the technology, its ability to burn less pure hydrogen than fuel cells, suitability for refit/retrofitting into existing vessels, and standalone nature of the power plant make ICE a more compatible power course with existing crew boat designs.

3.1 (c) Dual-fuel H2 Engines

Current dual-fuel engines are essentially conventional well-proven diesel engines, that are modified to aspirate hydrogen gas into the air intake. Combustion of the hydrogen inside the cylinder is then initiated by compression ignition, using diesel as a pilot fuel. The amount of diesel required to sustain combustion can be as low as 5%. In the event of issues with the hydrogen delivery system, or when the hydrogen fuel supply is exhausted, the engine can revert to running on diesel-only.

Although the availability of marine H2-ICE is scarce, a small number of dual-fuel engines are now on the market, with class approval. These engines are mostly medium or low-speed engines, however, a high-speed dual-fuel engine reached the market in mid-2022.

Anglo Belgium Company produces the Medium Speed DZ series with dual fuel capability or spark ignited for pure H2 combustion. The dual-fuel version can be fitted with an SCR catalyst and DPF filter for IMO Tier III emissions compliance when operating in diesel-only mode. As an illustration, the 8DZ model produces 1335 kW, with a dry weight of 13.9 metric tons and 4.8x1.6x3.1m (L x W x H) overall dimensions.

MAN diesel provides the much more compact high-speed D2862 series with dual-fuel capability. An integrated SCR system is also available for IMO Tier III emissions compliance. The LE448 model produces 749 kW of BHP, with a dry weight of 2.27 metric tons and 1.94x1.15x1.36m (L x W x H) overall dimensions. MAN reports that more powerful variants are currently in development.

Table 3. H2 dual-fuel engines for marine applications examples

	Power *(kW)*	**Weight** *(kg)*	**Dimensions** *(LxWxH) m*	**IMO Tier**
ABC 8DZ	1335	13900	4.8 x 1.6 x 3.1	Tier III (diesel only, with SCR and DPF)
MAN D2862 Dual Fuel	749	2270	1.94 x 1.15 x 1.36	

Current implementations of dual-fuel H2 engines are available only for the lowest power ratings, providing an unfavourable comparison of the power-to-weight ratio against diesel engines. Despite this, their current availability and relatively well-developed high technology readiness, and similarity with current FCB's design and operation philosophy make the dual-fuel engine an appealing prospect for adoption into FCBs, albeit currently at a slightly lower speed. Considering the removal of upward pressure on vessel speed previously discussed, this however is also an opportunity to offer further fuel savings.

3.2 H2 STORAGE OPTIONS

The following containment and storage technologies are considered in this paper:

- Compressed hydrogen gas: Hydrogen can be stored as a compressed gas in high-pressure tanks. This method is relatively simple and inexpensive but requires heavy and bulky tanks to achieve adequate storage capacity.
- Liquid hydrogen: Hydrogen can be cooled to a temperature of -253°C to become liquid. This method offers higher energy density than compressed gas but requires more complex storage tanks and equipment.
- Metal hydrides: Hydrogen can be stored in solid form within metal hydrides. This method offers comparable levels of storage density to compressed hydrogen for off-the-shelf variants but requires high temperatures and pressures to release the hydrogen.
- Liquid Organic Hydrogen Carriers: Hydrogen can be stored in liquid form at ambient conditions, with storage density comparable to liquid hydrogen. However, this technology is low TRL and as such not considered in the current work.

3.2 (a) Metal Hydride

In Metal Hydride storage solutions, hydrogen bonds to a host metal alloy and is stored in a non-reactive configuration. The system operates at a minimum temperature of 20 °C and it needs to be heated up to at least 60°C to release H2 gas. Since hydrogen desorption is governed by the physio-chemical process, no hydrogen will be released unless heat is added, thus reducing the risks associated with the hydrogen system. However, they are pyrophoric (reactive with oxygen) and/or reactive with water. The system provides an 88% overall efficiency, due to losses associated with the control processes during storing and delivery of hydrogen to the consumer equipment.

While metal hydrides are generally safer than compressed and liquefied hydrogen storage options, they release hydrogen slowly, making them less suitable for applications that require the rapid release of large amounts of hydrogen. They can also be more expensive than other hydrogen storage options, due to the cost of the metal hydride material and the need for specialised storage tanks and equipment. As such, metal hydride containment is not currently an attractive storage method for hydrogen in crew boat applications.

3.2 (b) Compressed H_2 Storage Tanks

Compressed hydrogen storage is generally less complex than liquefied hydrogen storage and does not require special equipment to maintain the cryogenic temperatures required to keep hydrogen in liquid form. Compressed hydrogen can be stored in tanks at 250 – 700 bar, as pressure increases so does the density of hydrogen (reaching 42 kg/m^3 at 700 bar). However, at higher pressures the bunkering process becomes more complex, requiring more sophisticated refuelling equipment, pre-chilling the gas to -40°C and careful control over the flow of hydrogen, thus increasing the operational costs[1].

While the volumetric capacity increases as a function of storage pressure the same does not occur for the gravimetric capacity. Increasing the working pressure also increases the structural requirements of the cylinders that need to be further reinforced. This drives an increase in the storage structure mass, which offset the benefit of packing more hydrogen in the same volume. For that same reason, compressed hydrogen storage solutions generally comprise several small diameter cylindrical pressure containers. These systems are typically deployed on modular containers that can be easily integrated into the vessel, bunkering and engine feed systems.

Commercially available solutions for compressed hydrogen storage onboard ships involve the use of Type IV storage tanks, made of composite materials, and usually offered in containerised form that can be easily integrated into the vessel as well as bunkering and engine feed systems.

A promising solution for crew boats is a containerised hydrogen storage solution offered by Marine Service Noord, with each 20ft container containing 400kg stored at 300 bar.

3.2 (c) Liquefied H_2 Storage Tanks

Liquefied hydrogen storage offers the best energy density, with liquid hydrogen reaching a density of 71 kg/m^3. These storage systems are similar to those used for LNG storage. They contain an integrated Tank Connection Space (TCS) with all the necessary equipment for safety, heating, and vaporisation to feed the power plant with hydrogen at regulated pressure and temperature. These storage systems need to be perfectly insulated to maintain the ultra-low temperatures of liquid hydrogen. This is achieved through a double-layer tank architecture, where the liquid is stored inside a steel tank, surrounded by an outer metal liner with vacuum insulation in-between. The double layer of the tank significantly increases the structural mass, although the resulting energy density is still higher than other hydrogen storage mediums. However, there is always a small heat leak from the ambient temperature outer liner through the insulating layer, through the inner metal liner and eventually to the liquid hydrogen itself. Heat leak to the liquid hydrogen causes boiling, with a build-up of hydrogen pressure within the inner tank. This boil-off is designed to match the consumption rate of the hydrogen, with sufficient additional volume in the tank to allow an adequate holding time. Typical boil-off rates depend on factors such as tank size and shape, and ambient temperature but are usually around 1% per day. Typically, these systems are designed for a 15 holding days requirement.

A handful of manufacturers offer commercial solutions for liquefied hydrogen storage onboard ships, however, most of them do not offer off-the-shelf solutions, but rather custom-built, depending on the requirements of each vessel.

MAN Cryo currently offer the most viable liquified H2 storage solution for crew boats with a Class-approved system with a gross volume of 50 m^3 that can hold up to 2.3 t of net usable liquid hydrogen.

3.2 (d) H2 Storage Summary

The H2 storage technology utilised will have a significant impact on the bunkering process, an important aspect of the operational viability of the vessel. Both compressed H2 and metal hydride allow the sourcing of compressed H2 while liquid H2 storage systems require the sourcing of cryogenic H2. The bunkering process for cryogenic H2 has additional requirements to prevent heat leakage and the availability of cryogenic H2 is severely limited outside of European territories. It should also be noted that the availability of green hydrogen is currently poor and therefore it is likely that blue hydrogen would be used in the near term unless it is made in situ to support a specific operation, similar to the Torghatten ferry or EMEC in Orkney.

A comparison between the different technologies presented herein is shown in Table 4.

Table 4. Traffic light qualitative comparison of storage technologies

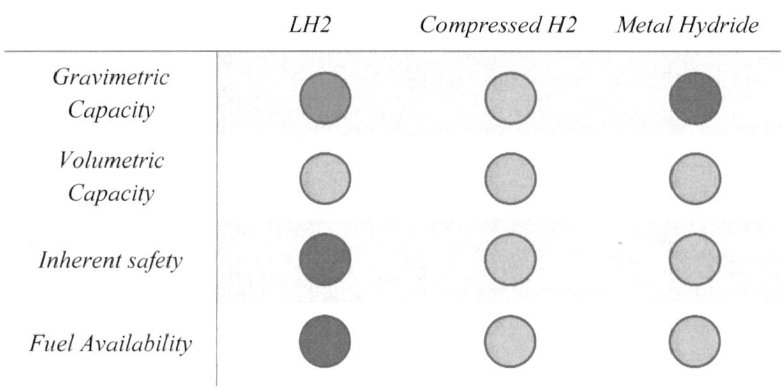

From Table 4, it can be seen that compressed H2 offers a good compromise across all compared factors. In particular, the limited availability of Liquid H2 makes it unsuitable for use on commercial vessels with high yearly operating hours such as crew boats. Furthermore, the low gravimetric capacities of the metal hydride solution make this solution undesirable for a low displacement high-speed vessel.

As such, comparative analyses undertaken in the following sections of this paper are based on the use of a compressed H2 containment system.

4. REGULATORY REQUIREMENTS IMPACT ON GENERAL ARRANGEMENT

Since the regulatory landscape for H2 storage and propulsion is not yet fully developed, statutory approval is given through risk-based certification, which consists of the identification of risks and successful mitigation of them. This process provides some instability in the design requirements depending on the project and storage technology.

In broad strokes, the major impacts on the GA are driven by the fire and leak risks. There are several features, taken from the IGF code for LNG, to reduce the probability and impact of such risk. Hazardous areas are classified as locations that can routinely or exceptionally (i.e. due to a fault or failure) contain an H2 atmosphere. The IGF code provides the classification of hazardous zones at varying radii of potential H2 emission locations. This is a preliminary estimate that would need a dispersion simulation to be validated.

Inside a hazardous zone, all electrical equipment needs to be explosion-proof. Electrical wiring and equipment are generally prohibited unless essential to operations in that area. Care should be taken not to overlap such zones with working areas and entrances to non-hazardous internal compartments.

Internal compartments that contain hydrogen systems need additional structural fire insulation, blast protection, gas-tight arrangements, fire extinguishing systems, fire main specific configurations and adequate ventilation. They must also be treated as emergency shutdown protected machinery spaces, such that in an emergency the vessel must be able to shut down any non-safe equipment in that area. There are several methods to mitigate the accumulation of a flammable cloud of hydrogen, for example, a pressure gradient between compartments, blast-proof ducting and ventilation at a minimum of 30 air changes per hour. To that end, any entrance to that compartment must have an airlock to minimise the risk of hydrogen entering a non-hazardous area. Any ventilation outlets should also be treated as a source of hydrogen atmosphere and their associated hazardous area should not overlap with ventilation inlets. Placing the vent as high as possible on the ship mitigates inlet overlap and potentially flammable external regions. All piping for H2 will need to be double-walled to mitigate the potential for a pipe to break and leak hydrogen.

Compressed hydrogen cylinders will be fitted with pressure relief valves (PRVs), such that the tanks can be emptied within a given timeframe to prevent the tank from rupturing due to pressure accumulation. In the event of the PRVs working the hydrogen should be vented in such a way as to minimise the accumulation of the flammable cloud. A vent mast must also be included for the case of an emergency, controlled H2 discharge, however, this has minimal impact on the layout and cargo deck capacity. Means of purging storage vessels and piping should also be included, as well as additional fire protection measures. These are a few of the requirements that one can expect to be driven by the risk assessment process in a design evaluation class approval. Therefore, it is fair to expect a significant increase in the vessel lightship by including an H2 storage system.

5. H2-FUELLED CREW BOAT CONCEPT

5.1 FEASIBLE CONCEPT DESIGN

For application in a crew boat design, the availability of dual-fuel engines, compatibility with current crew boat designs and the possibility for refit, lead to the dual-fuel engine being considered the most viable hydrogen-fuelled power plant. Compressed hydrogen is considered the most viable and practical choice of hydrogen storage/containment for this application, as discussed in Section 3.2.

Although crew boats have typically been designed for a speed approaching 30 knots, the reduced speed pressures discussed in Section 2 lead to a less powerful vessel being commercially viable. Consequently, although the basis design for this study is capable of 30 knots, with an installed power above 3200kW, the dual-fuel variant has a lower installed power and speed. The combination of dual-fuel operation, with a lower service speed is expected to lead to significant reductions in diesel fuel consumption and by extension emissions.

As a compact high-speed engine, the MAN D2862 presents few challenges to integrate into existing crew boat engine rooms, although modifications will be required to handle hydrogen, which typically uses high-speed engines such as the conventional diesel-fuelled MAN D2862 LE483 and CAT C32. The dimensions, weight and power output allow its use as a drop-in replacement in typical engine room arrangements, opening the possibility of installing dual-fuel capability into existing vessels during refit. While other considerations are necessary for the operation of a hydrogen-powered engine, the availability of high-speed dual-fuel engines critically means that the engine change does not require a change in vessel length (as would be the case with a medium-speed dual-fuel engine).

An on-deck containerised compressed hydrogen storage system is selected as the lowest impact storage solution on the existing design and also aligns with the possibility of retrofitting dual-fuel engines. Although this solution reduces the available cargo deck space, placing the fuel on the deck substantially reduces the engineering and risk mitigation complexities of storing hydrogen below deck. Additionally, locating the hydrogen fuel below the deck would necessitate a vessel length increase from 42m to a minimum 45m to accommodate the same level of internal net space.

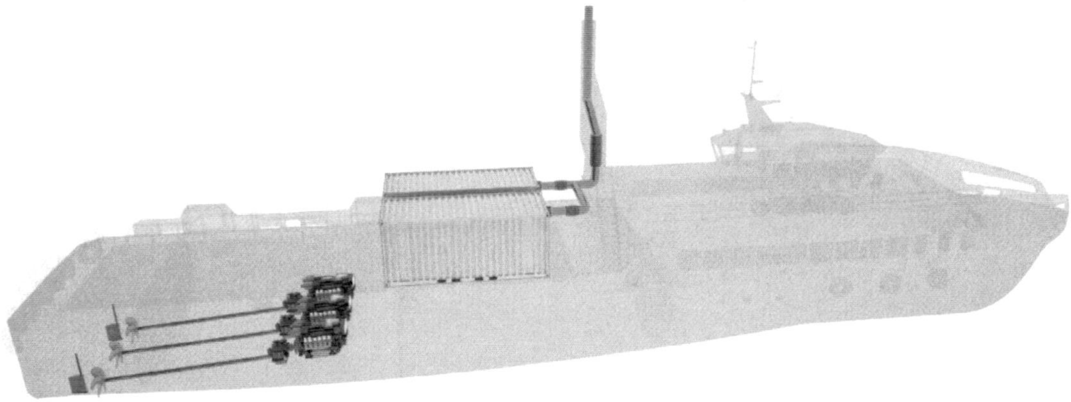

Figure 2: 2 x 20ft containerised on-deck compressed H2 storage, with H2 high-speed dual-fuel engines

Two dual-fuel variants are considered, featuring either one or two on-deck compressed H2 storage units. Each container has a 400 kg capacity of compressed H2, giving each variant a different range when operating in dual-fuel mode.

Considering the fuel storage capacity selected for this design, the vessel will be able to operate on dual-fuel mode, at the recommended energy balance a for limited time, after which the vessel should revert to diesel propulsion until all remaining diesel has been consumed.

The performance of four variants is presented here, as follows:

- Design 1 - Basis of design: a well-proven 42m crew boat, with a top speed of 30 knots, powered by conventional diesel-mechanical (D-M) engines
- Design 2 - Conventional diesel-mechanical 42m crew boat, with less installed power, achieving approximately the same top speed as the dual-fuel designs
- Design 3 – Dual-fuel variant 42m vessel with 1 x compressed H2 storage unit (400kg H2)
- Design 4 – Dual-fuel variant 42m vessel with 2 x compressed H2 storage unit (800kg H2)

Design 1 shall be used as the baseline vessel, against which the performance of other variants is measured. As much of the analysis is based on operational profiles that use engine MCR as their basis, Design 2 is included in the analysis to allow conclusions to be drawn about whether the fuel-saving shown is a result of the reduced installed power, or as a result of the dual-fuel operation.

Table 5 shows the comparative principal particulars of these variants:

Table 5. Principle Particulars of design variants

	Design 1 42m D-M FCB	**Design 2 D-M FCB with lower top speed**	**Design 3 Dual-fuel FCB w/ 400kg H2**	**Design 4 Dual-fuel FCB w/ 800kg H2**
Length (m)	42.0	42.0	42.0	42.0
Lightship Displacement (tonnes)	132.0	127.6	140.4	149.1
Engine Make/Model	CAT C32	MAN D2868 Dual-fuel	MAN D2862 Dual-fuel	MAN D2862 Dual-fuel
Engine Power (kW)	3 x 1081kW	3 x 662W	3 x 749kW	3 x 749kW
Top Speed (kn)	28.5	22.5	22.7	21.7
Working Deck Area (m2)	110	110	90	62

Available Deadweight (tonnes)	148	150.4	139.3	130.9
Pax	80	80	80	80
H2 Capacity	-	-	400kg	800kg

5.2 PERFORMANCE ANALYSIS

The capabilities and performance characteristics of the crew boat variants discussed in Section 5.1 has been evaluated, using Option 1, the well-proven 42m diesel-mechanical variant, as the baseline for comparison.

When fuelled with the maximum design capacity of H2, the range of each of the dual-fuel variants when operating in dual-fuel mode (assuming H2 to be the limiting fuel) is shown in Figure 3. The increase in range above 21 knots is driven by the significant shift in MAN's recommended energy balance towards diesel consumption at higher MCRs. This is seen in the increase in diesel consumption at this speed.

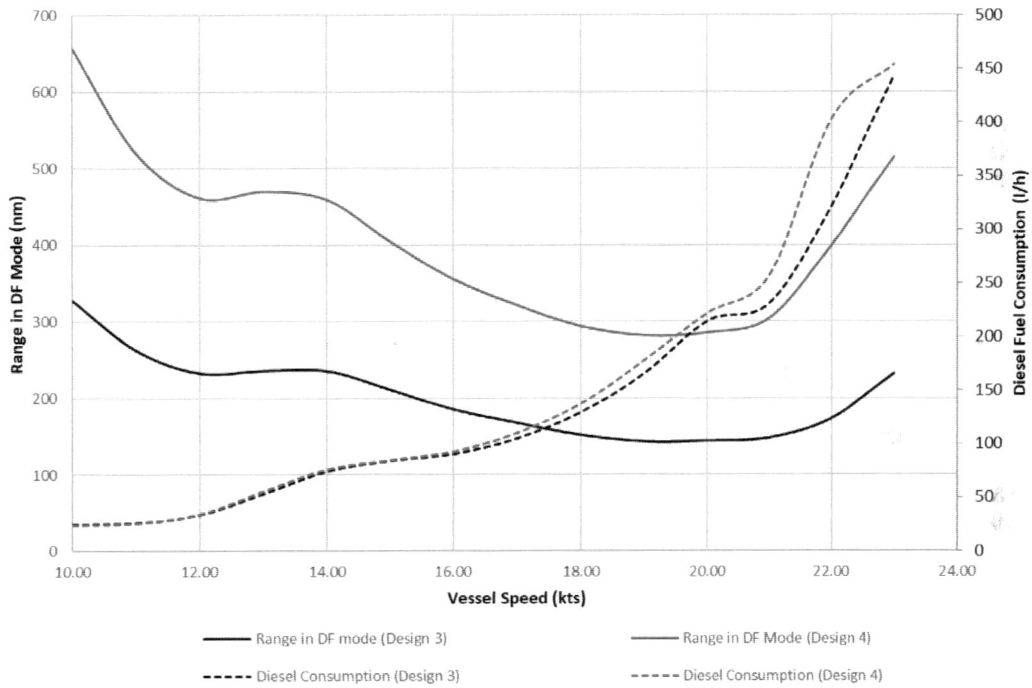

Figure 3: Speed vs. Range and Fuel Consumption when operating in dual-fuel mode

Showing diesel fuel consumption as a percentage of the baseline (Design 1) reveals significant reductions in fuel consumed when operating in DF mode, most prominently at lower speeds where reductions in diesel fuel consumption are up to 60%. Due to a slightly lighter displacement, Design 2 also shows a small reduction in fuel consumption at lower speeds.

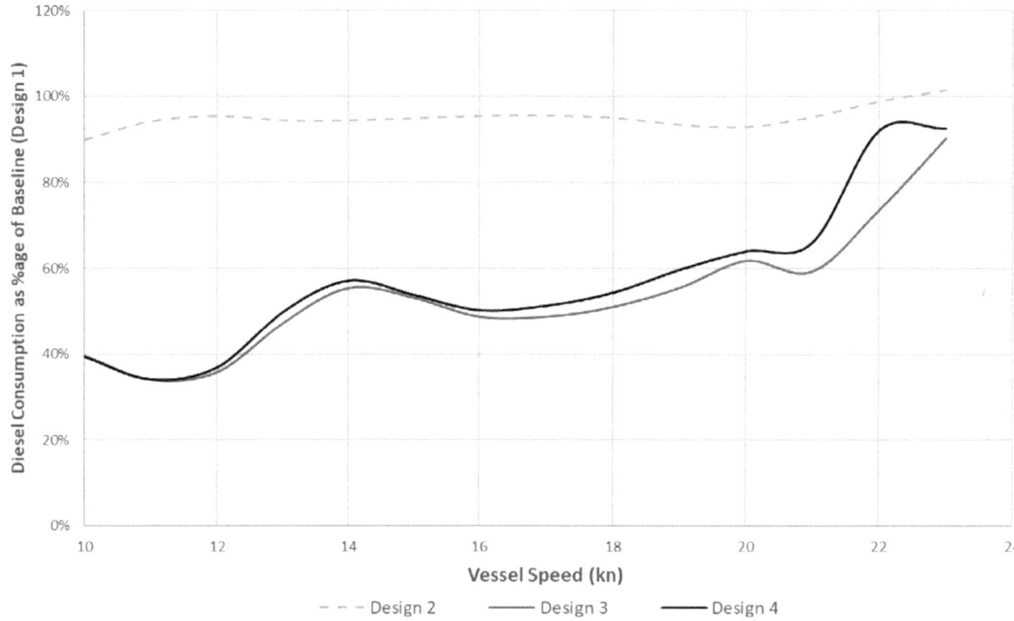

Figure 4: Speed vs. Fuel Consumption as a proportion of the baseline fuel consumption (Design 1)

It is expected that the performance benefit of dual-fuel vessels will vary with differing operational load profiles. Consequently, 4 load profiles are considered in this analysis;

- Profile 1: Light running load (low average MCR %age)
- Profile 2: Heavy running load (high average MCR %age)
- Profile 3: Sample MCR profile recorded from a 3-day track of a vessel operating in the Persian Gulf
- Profile 4: Sample MCR profile recorded from a 3-day track of a vessel operating around Singapore

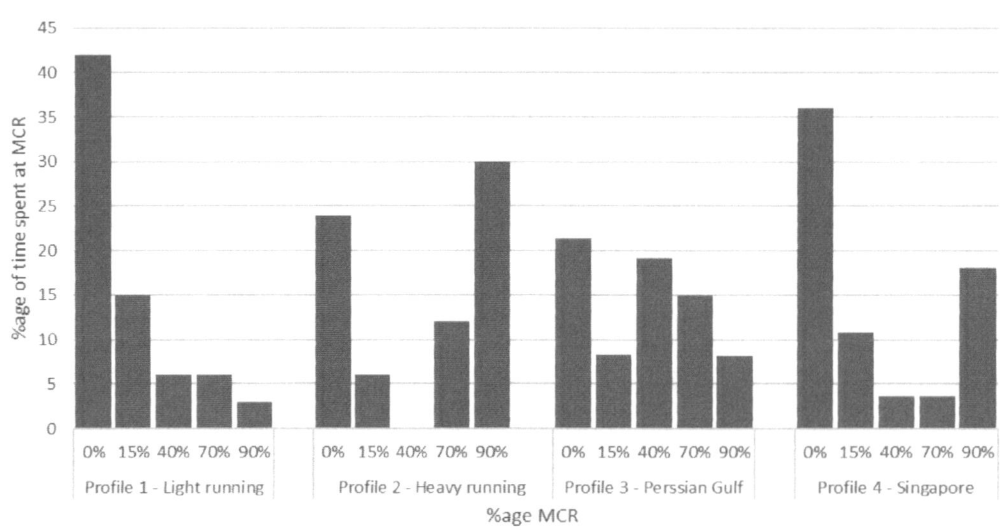

Figure 5: MCR profiles considered in concept performance analysis

A 3-day mission is analysed, using each of the load profiles above. It is assumed that vessels will use their reserve of H2 when operating in dual-fuel mode at lower MCR operations and will expend all H2 during the mission. The remainder of the operations will be conducted in diesel-only mode. A fixed fuel load and deadweight for each variant are considered. Dual-fuel variants are assumed to operate at the supplier's recommended energy balance.

As all variants require a generator to provide an electrical hotel load, and the fuel consumption associated with this will be approximately equal for all variants, fuel consumed by auxiliary generators is not included in the figures presented here.

Diesel consumption during a 3-day mission of each load profile is shown in Figure 6. With a finite supply of H2 onboard, which is not sufficient to complete the entire mission in DF mode, part of the profile is completed in DF mode, with the remainder being completed as a pure diesel-driven vessel. As diesel consumption reduction in DF mode is greatest at low MCRs (see Figure 4), it is expected that it is most efficient to consume H2 reserves during the lower MCR phase of the load profile; the subsequent analysis, therefore, makes this assumption.

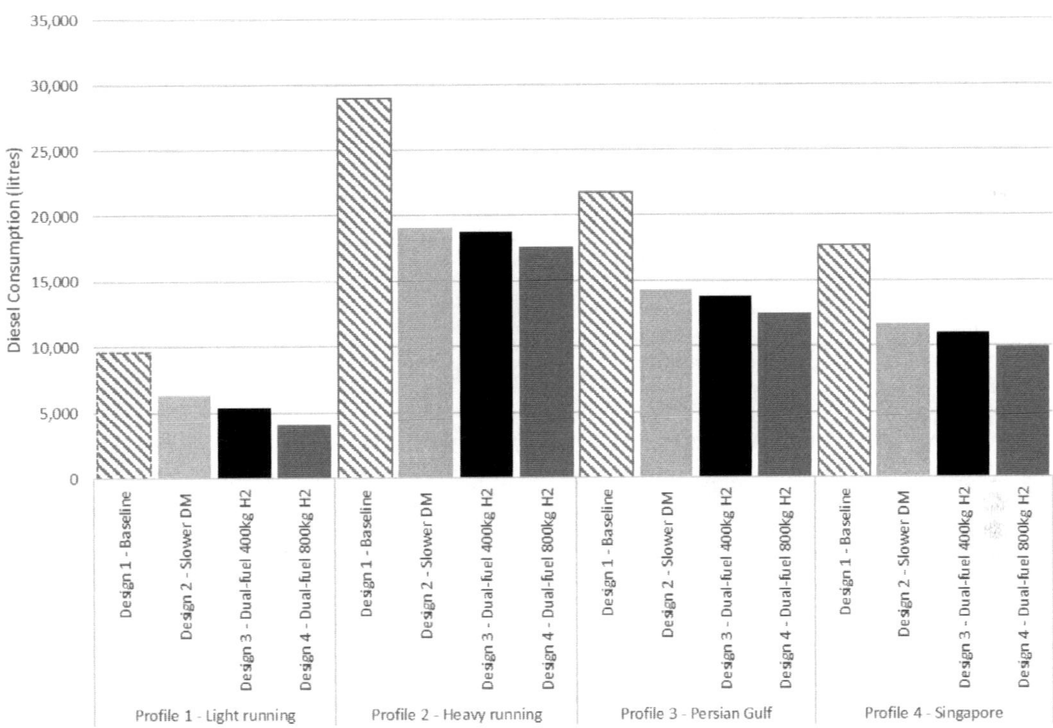

Figure 6: Diesel consumption of each design during 3-day mission of varying MCR profiles

Figure 6 shows a significant reduction in fuel consumption of all alternative designs over the baseline design, in all operational profiles, with diesel fuel savings approaching 60% in light-running operating profiles. This is principally a result of the reduced operating speed, however, there are meaningful reductions in diesel consumption (up to 36%) of the dual-fuel variants, when compared to the lower-powered diesel-mechanical vessel (Design 2).

Figure 7 shows the savings in diesel as a %age of consumption of Design 2, the slower diesel-mechanical vessel. Fuel savings are most prominent in lighter running operational profiles; this is to be expected from Figure 4, over 35%.

It is important to note these figures represent a 3-day mission, with no refuelling mid-mission. If operational schedules permit more frequent refuelling of H2 the proportion of the mission operated in dual-fuel mode would increase, realising further reductions in diesel consumption. An alternative approach is to increase installed H2 storage capacity, however, this comes with additional lightship weight of the containment system and reduced deck area.

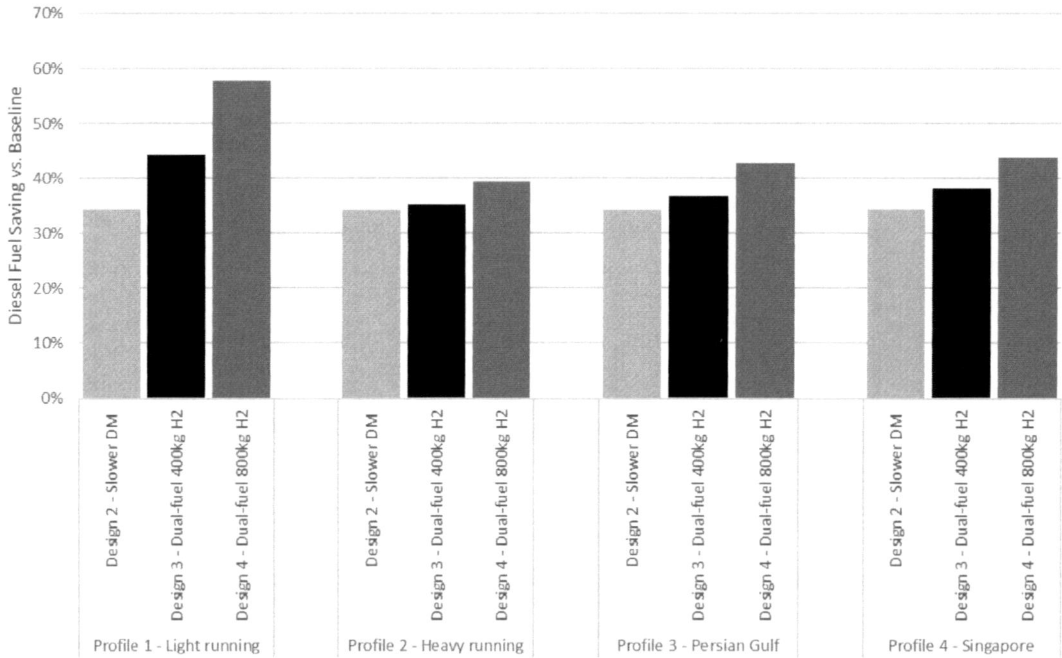

Figure 7: Diesel fuel consumption savings during 3-day mission as a percentage of Baseline (Design 1)

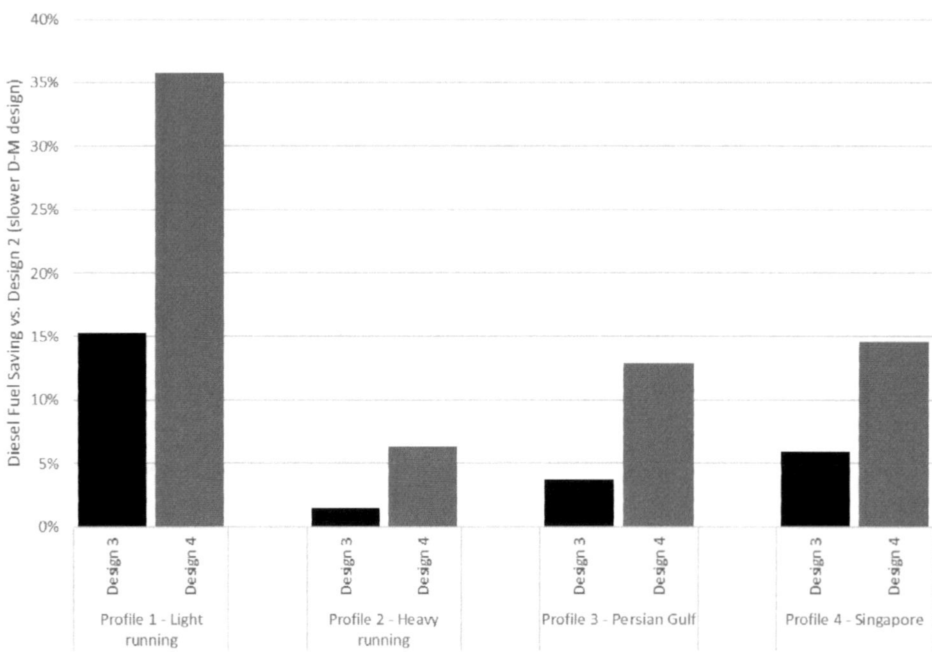

Figure 8: Diesel fuel consumption savings vs. Design 2 (slower diesel-mechanical design)

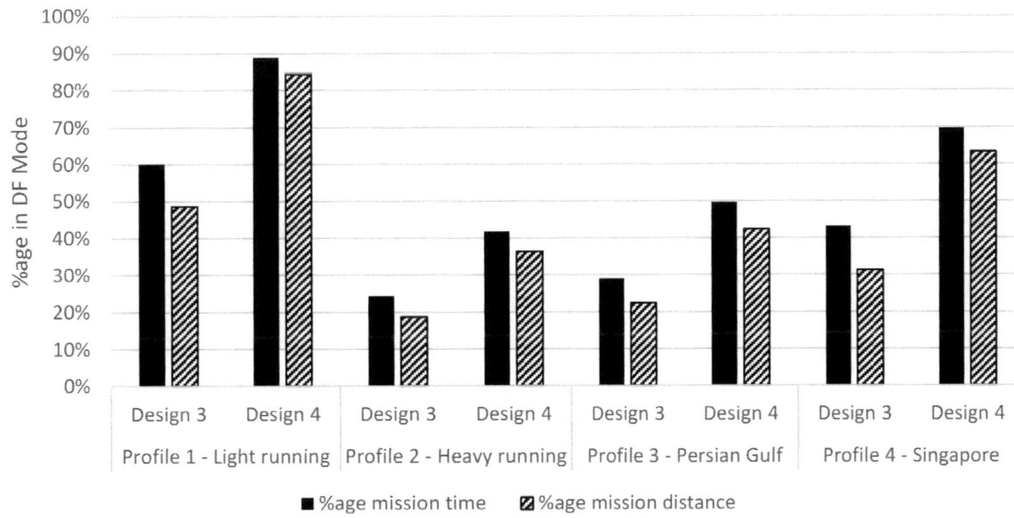

Figure 9: %age of operation in dual-fuel mode

The quantity of hydrogen stored has a significant impact on the reduction of diesel, as shown in Figure 8, however, comes at the expense of available deck area. This reduction in diesel consumption is directly proportional to the tank-to-wake reduction in CO2. However, for a holistic view of carbon emissions, well-to-wake emissions are to be considered which would be heavily reliant on the production method of the hydrogen. The tank-to-wake emissions would be directly comparable if the hydrogen used is from renewable power (green hydrogen) and would be similar if hydrogen production utilises carbon-capture (blue hydrogen). Given that availability of blue or green hydrogen is currently very limited, in the immediate future it is unlikely to be the H2 fuel source for dual-fuel crew boats.

While conclusions can be drawn around the overall viability of the propulsion system and general benefits in fuel consumption reduction, the variance between route profiles shown in Figures 7, 8 and 9 also illustrates the importance of considering the vessel's specific operational profile in assessing the benefits for a particular vessel.

6. POTENTIAL DEVELOPMENTS

This article has investigated the potential of using dual-fuel engines in the context of a crew boat design. As is typical with ship design, there are always alternative ways of solving the same problem or providing similar capabilities. This section highlights a few potential modifications to the proposed design.

One alternative would be to increase the vessel length from 42m to 45m and place the H2 storage system under the deck. This arrangement would require resolving the additional complexities in arrangement and fire safety while having H2 storage in an enclosed compartment, next to the engine room and crew cabins. However, this arrangement would allow the vessel to retain its entire deck cargo capacity, greatly improving its operational capabilities. Such a vessel could be used as a seamless substitute for existing 42m crew boat designs.

Another intriguing possibility lies in using the MAN high-speed engine block rated for a higher power. It is reported that MAN is currently pursuing certification of a 1450 kW engine model. This would allow the design of a vessel with similar or superior speed capabilities to the reference 42m diesel-only Fast Crew Boat Vessel.

The concept presented here focuses on the conversion of an existing design to incorporate H2 and reduce emissions and fuel consumption with a design concept that will be familiar in terms of technical operation by most operators. However the blank-paper design exercise of an H2 vessel, without the limitations of operator familiarity would likely yield a different design, possibly utilising the higher inherent efficiencies of fuel cells.

Finally, the longer-term viability of dual-fuel engines should be considered as society moves towards net-zero emissions. This paper shows that dual-fuel engines are technically viable to use in FCB designs of today, and may provide cost and fuel savings to operators depending on the relative cost of installation, maintenance and fuels, and the operator's requirements and profile. However, dual-fuelled engines should not be the end-point in development and must be viewed as part of an energy transition. The logical net-zero endpoint for dual-fuelled vessels is the installation of a full H2-ICE engine, either through design evolution or engine replacement during the vessel's life. However, the quantity of H2 necessary to achieve a range and endurance equivalent to today's diesel-powered FCB is likely to be prohibitive for a 42m FCB. As such, dual-fuel engines in crew boats should be viewed as a method of reducing the carbon footprint, but not a route to net zero.

7. CONCLUSIONS

Hydrogen storage is a challenging problem with different solutions available to the designer, which vary greatly in terms of performance, availability and operational requirements. Several storage options have been considered with compressed H2 selected as the most viable option for existing crew boat designs because of its ease of sourcing around the world and relatively simple integration into existing designs. Of the H2 propulsion systems considered, dual-fuel engines offer an attractive combination of relatively high TRL, mature technology base, resilience, and compatibility with current designs. As such, the combination of dual-fuel high speed engines, along with on-deck compressed H2 storage is considered to be the best currently available technology for hydrogen integration into crew boats, to provide fuel diversity and resilience, and reduce diesel consumption and emissions.

The dual-fuel crew boat has been demonstrated as a technically feasible platform with reduced diesel-fuel consumption, as compared to a comparable diesel-mechanical variant. The proposed implementation limits the overall impact on existing designs. The quantity of H2 carried has a significant impact on the overall diesel reduction, as it increases the proportion of the mission that the vessel can operate in dual-fuel mode.

The concept design assumes a lower maximum speed than the existing baseline design as a direct result of the limited power of dual-fuel engines currently available. There are inherent fuel savings available by reducing the operating speed. However even when comparing the dual-fuel vessel to a diesel-mechanical crew boat, with a comparable design speed, diesel savings of up to 35% are seen in the operational profiles studied.

It is found that consuming the H2 supply at the lower MCR parts of the operating profile rather than higher MCR parts offers the greatest overall reduction in mission diesel consumption. Refuelling interval is an important factor in the amount of diesel saved: more frequent refuelling of H2 increases the proportion of time that it is possible to run in dual-fuel mode, and increases diesel saving. Some of the deck area is lost to H2 storage, which could be compensated for with a slight increase in vessel length. Data on fuel savings is presented to facilitate further discussion and cost-benefit analysis.

8. REFERENCES

1. MARINHO, T., 'Dual Fuel Crew Boat Technical Note', *BMT*, 2022

2. Office of Renewable Energy, 'Hydrogen Storage', *https://www.energy.gov/eere/fuelcells/hydrogen-storage* (Accessed: 04/05/2023)

3. HOECKE, L V., et al, 'Challenges in the use of hydrogen for maritime applications, *Environment and Energy Science*, 2021

4. HYDE, K., and ELLIS, A., 'Feasibility of Hydrogen Bunkering', *Interreg North Sea Region Programme*, 2019

5. FESMIRE, J., et al., 'Energy efficient large-scale storage of liquid hydrogen', *IOP Conference Series: Materials Science and Engineering, 1240, 012088*, 2022

9. AUTHORS' BIOGRAPHY

James Calver holds the current position of Principal Naval Architect at BMT SSD with experience in the concept, class, and production level design of a range of commercial vessels. James leads the Naval Architecture team across the varied portfolio of BMT SSD, which has an established track record in the design of Commercial, Offshore Energy and Government and Security vessels.

Christophe Rident is the Lead Naval Architect for Commercial Vessels and High-Speed Ferries at BMT SSD, with extensive experience in designing vessels using alternative marine propulsion such as electric and hybrid systems, and alternative fuels.

Dr. Thomas Beard holds the current position of Alternative Fuels Lead at BMT UK and has a Doctorate in Hydrogen Safety. Thomas has experience in alternative fuel safety and integration into a variety of vessels for both new-build and retrofit.

PROACTIVE SHOCK MITIGATION: REDUCING INJURY AND IMPROVING PERFORMANCE OF CRAFT AND CREW

J Salmon, Salmon Power Sports, UK, **B Landgrén,** PerformanSea, Spain

SUMMARY

Dynamic ballast systems have been used for decades by Search and Rescue (SAR) organisations such as the Royal National Lifeboat Association (RNLI), and are proven to reduce vertical acceleration levels. Dynamic ballast can therefore be utilised to reduce the risk of injury onboard High-Speed Craft (HSC), and to improve the effectiveness of crews in performing their assigned tasks.

However, unlike external trim-control systems such as tabs or fins, ballast systems are inherently complicated to retrofit into existing craft, and should therefore be considered from the onset of the design stage. This simulation model has been developed to assist naval architects who may be unfamiliar with the technology. Here the effect of dynamic ballast upon a 13m Rigid Inflatable Boat (RIB) is studied, noting the relation between pure trim adjustment versus the addition of mass, the speed of travel, as well as comparing effects at different hull deadrise angles.

1. INTRODUCTION

Many professional boat crews must, by necessity, operate in rough seas. Rescue services are inherently likely to be called upon when conditions are poor, whilst time-critical military operations cannot be postponed for calmer seas. However, high-speed transit in rough seas can result in significant impact events [1], where the vessel is slammed against the water surface. The nature of the vertical acceleration may change depending on the design of the craft and the shape, size, and frequency of waves [2], but is typically in the form of an angular motion, with the bow of the craft being raised prior to impact. This shock travels throughout the craft, reaching the passengers and crews onboard and subjecting them to pain, potential injury, and fatigue [3]. As well as the risk to humans onboard, extreme levels of impact can put the entire craft at risk of structural failure [4].

1.1 EFFECTS OF IMPACT

The most commonly reported injuries are to the neck, lower back, shoulders, and knees [1], and there is also evidence for neurological damage caused by frequent high-level impacts. Crews also report that these conditions are detrimental to mental acuity and physical capability [5]; in a retroactive study [6] of combat crewmen, 70% felt that they had experienced an impaired capacity to perform their assigned roles due to impact and Whole Body Vibration (WBV) exposure.

Whilst these injuries may be expected during operations in rough seas, the risk remains even in calmer conditions. One study [7] of HSC injuries reported that 61.5% of "deck-slap" injuries occurred during transit in calm seas, due to the reduced level of caution demonstrated by the helm.

Table 1. Shock mitigation techniques

Category	Method
Design stage	Human-factored design Hull form
Behavioural	Specialist training Intelligent guidance systems Governmental guidance and regulation
Pitching reduction	Dynamic ballast Trim tabs Interceptor fins
Personal protection	Suspension seating Exoskeletal support

Given that HSC transits can be the initial stage of critical military [8] or lifesaving operations, it is imperative to reduce these injurious and performance-limiting effects as much as possible.

1.2 MITIGATING IMPACT

Much of the existing focus on the protection of crews against slamming has centred on minimising the transfer of shock from the craft to those onboard, with shock-mitigating suspension seating now relatively commonplace onboard many professionally operated HSC. Whilst suspension seats have proven an effective option for reducing shock loads to individual occupants [9], a multi-level holistic approach to shock mitigation is required [10] to protect crews against injury. Examples of potential methods are listed in Table 1.

Whilst each method of shock mitigation brings its own merits to a combined overall approach, a specific advantage of the targeted reduction of pitching motions is that the benefits are applied to all occupants, as well as the craft itself. Decisions can be made during the design stage in order to optimise hull form for the intended application; for example, higher deadrise angles typically result in reduced levels of vertical acceleration at the bow [11], but experience higher running resistance. Designers may also reduce pitching motions through the addition of trim control systems such as trim tabs, interceptors, or dynamic ballast.

1.3 DYNAMIC BALLAST

The use of dynamic ballast systems on HSC originated in offshore powerboat racing, as a means of adjusting trim angle without compromising speed through the use of drag-creating [12] external appendages such as trim tabs, and has since been adopted by SAR organisations for use in rough-water operations. By taking on sea water into a forward ballast tank as required, craft displacement can be increased and Longitudinal Centre of Gravity (LCG) shifted forwards in order to increase inertial resistance, thereby reducing slamming. Tanks are most commonly located at the bow of the boat; as the largest accelerations typically occur there, and positioning the additional mass as far forwards as possible makes the most of the effective leverage.

Sea trials of a 7.5m lifeboat demonstrated that a ballast tank equal to 12% of the craft's original displacement, resulted in an approximately 50% reduction of average vertical acceleration levels, with an approximately 70% reduction of peak acceleration levels [13].

Table 2. Dynamic model inputs and outputs

Category	Inputs	Outputs
Boat	Length of waterline Chine beam Deadrise Mass (loaded) LCG Forward velocity	Mass flow rate to/from tank Average vertical accelerations Dynamic trim angle Hull resistance
Ballast system	Tank cross-sectional area Tank height Vertical location of tank Longitudinal location of tank	
Environmental conditions	Significant wave height	

Dynamic ballast is also proven to be effective on larger craft; in one study [14], simulations compared a 24m aluminium naval patrol craft against an identical design featuring a lightweight carbon fibre hull construction. The structural mass of the carbon craft was approximately 50% lower than the aluminium version, with a total displacement reduction of 18%, corresponding to a 15% reduction in fuel consumption and CO_2 emissions. However, the lower displacement of the carbon craft also meant that vertical acceleration levels were increased by 20% compared to the heavier aluminium version. The addition of a forward ballast tank on the carbon craft, equal to 4% of the total displacement, conversely reduced vertical accelerations by approximately 30%, resulting in improved rough water performance compared to the aluminium craft. It should be noted however that aluminium and carbon fibre have very different structural properties and behaviours; the above example relates purely to the range of vertical acceleration experienced at the bow.

There are clear benefits to the use of dynamic ballast on HSC, however the internal nature of the system makes it inherently more complicated to retrofit onto an existing craft than external systems such as trim tabs. It is therefore required that naval architects consider the use of ballast from the early design stage.

2. DYNAMIC MODEL

The simulation model of the dynamic ballast system has been developed to assist with performance estimations and feasibility studies during the design stage, in a similar manner to the performance prediction calculations provided by propulsion systems manufacturers. The simulation model is modular, and consists of three blocks:

- Ballast tank and scoop dynamics
- Smooth water dynamics of hull
- Rough water performance of hull

2.1 LIMITATIONS

The model is intended to provide insights during the early design stages when only limited amounts of craft data are available, but critical decisions on how best to meet performance requirements are being made. Therefore, the model remains a simplification of a complex hydrodynamic problem, with empirical formulas that have strict boundaries of applicability. The vertical acceleration model is based on the established Savitsky & Brown model, which has also been adopted by several classification societies such as DNV and RINA.

Due to the complexity of the subject, the validation and expansion of the dynamic model is ongoing. The reporting of HSC vertical accelerations currently lacks a standardised form [15], which makes the comparison of data from different sources often impossible and can introduce significant errors, which can complicate the validation process. Another potential source of error lies in difficulty to predict dynamic trim angle which mirrors vertical acceleration in rough water [16]. It must therefore be noted that this simulation model will be continually developed as more data is made available and validation is made possible.

Table 3. Simulation parameters

Parameter	Unit	Simulation 1	Simulation 2
Length of waterline	m	12.2	12.2
Chine beam	m	2.9	2.9
Deadrise	°	22	20, 22.5, 25
Mass (loaded, excl ballast)	kg	10,000	10,000
LCG (excl. ballast)	m	4.27	4.27
Ballast tank volume	m^3	0…1.200	0…1.200
Ballast tank longitudinal location (from transom)	m	12.2	12.2
Significant wave height, H1/3	m	1	1
Beam-based Froude number	-	1.5, 2.0, 3.0	1.5
Forward velocity	m/s	8,10.7,16	8
Forward velocity	kts	15.6, 20.1, 31	15.6

2.2 INPUTS AND OUTPUTS

The most significant inputs and outputs of the model are listed in Table 2.

Although many HSC manufacturers make technical specifications of their craft publicly available, there is no standardised format; for example, some may provide an estimation of total weight including fuel and engines, whilst others may provide only lightship data with an indication of maximum engine horsepower. Different variations to measuring beam, LCG, and deadrise exist, which complicates the selection of input parameters for the simulation. Therefore, for the future validation of this model, a questionnaire has been developed in order to consistently acquire appropriate data for all required inputs.

2.3 SIMULATION

For testing purposes, an existing model 13m fibreglass hulled RIB was selected; the model was selected based on the suitability of available technical data, the stated intended usage of the craft, and as a representative midpoint of the normal size range for applicable craft.

For these simulations, the ballast tank was positioned as far forwards as possible in order to replicate an integrated bow tank structure. The tank utilises baffles to minimise potential instability caused by the Free-Surface Effect (FSE) when the tank is only partially filled, so accounting for FSE instability is included in this simulation. An allowance of mass was included to account for fuel and passengers, and a tank volume of 12% of total loaded mass was selected based on previous studies [13].

The speed range was selected so that it covers pre-planing, planing hump, and pure planing speeds. The speeds were non-dimensionised to beam-based Froude numbers (FNB) to allow for easy comparison. The input parameters for this simulation are outlined in Table 3.

Two simulation runs were made for this test of the model; in the first the speed was varied, whilst in the second the deadrise angle of the hull was varied.

3. RESULTS

Figure 1 shows the results of Simulation 1; the reduction in vertical accelerations (Acc_z) has been plotted against FNB and ballast tank volume (V_tank). To have a better understanding of the model's sensitivity, the effect of dynamic trim angle was plotted separately (upper surface) from the combined effect of trim angle and ballast mass (lower surface).

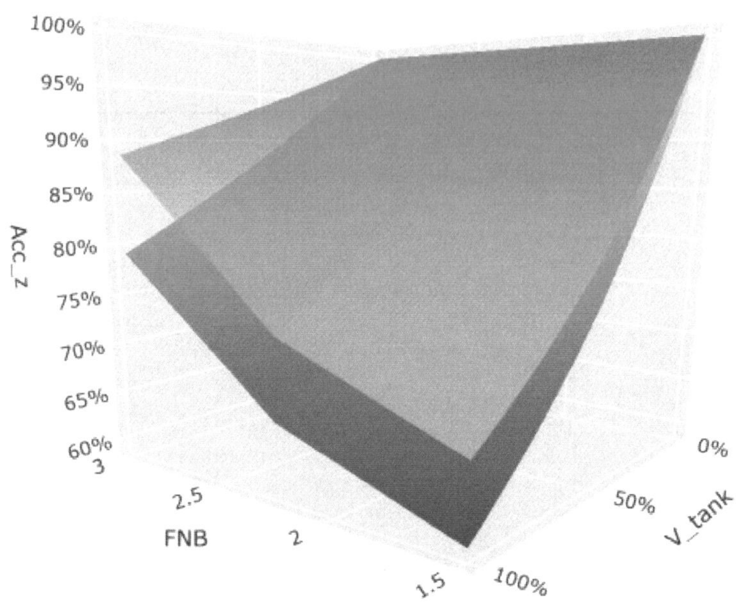

Figure 1: Simulation results, Beam-based Froude number

The increase in tank volume corresponds with a 22-43% decrease in vertical acceleration levels across different speeds.

Figure 2 shows the result of simulation 2, with the reduction in vertical acceleration levels (Acc_z) at the bow plotted against the deadrise angle (Beta) and ballast tank volume (V_tank).

4. DISCUSSION

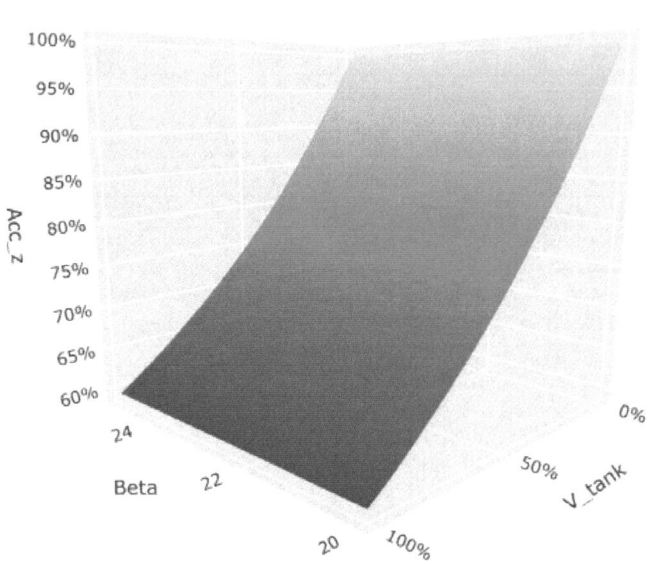

Figure 2: Simulation results, deadrise angle

4.1 MASS EFFECT

Due to the increased inertial resistance generated by the additional mass, the lower surface demonstrates consistently lower vertical acceleration levels compared to the upper surface showing the effect of trim only. At higher speeds, this difference as a proportion of overall potential acceleration reduction increases. For example, at FNB = 1.5 the effect of trim angle accounted for 75% of the total reduction of vertical acceleration. In other words, the effect of added mass accounted for 25% of the reduction in vertical accelerations.

However, as the speed was increased to FNB = 3, the effect of trim on vertical accelerations reduced. Here the trim and added mass played equal parts in reducing the vertical accelerations, suggesting that the "mass effect" gains importance as speed increases. The positive correlation between speed and the effectiveness of increasing mass makes sense, as the forces resisting the change of trim angle increase with speed.

This also emphasizes the significance of the dynamic trim angle in reducing vertical accelerations. By adjusting the LCG or generating additional lift, the dynamic trim angle can be altered. To simplify the comparison, we can consider the impact of extra lift as a shift in effective LCG (LCGe).

The LCGe depends on the extra lift, which can be attained by employing external devices like trim tabs and interceptor fins. The shift in the LCG on the other hand is influenced by the distribution of mass, which can be achieved through the utilization of a dynamic ballast system.

The shift in LCG is hence not affected by the speed of the craft or environmental conditions, and remains a constant effect upon the craft while the system is engaged. However, the magnitude of the LCGe varies with craft speed and is susceptible to disturbances like flow turbulence and water aeration, making it less reliable in heavy seas.

4.2 DEADRISE

Boats with steeper deadrise angles experience lower vertical accelerations, but typically have higher running resistance [11] resulting in higher operating costs. Conversely, boats with shallower deadrise angles are more efficient in calm water, but can experience high levels of impact in heavy seas. However, as highlighted in Figure 2, the model suggests that shifting LCG forward with ballast in the bow can reduce vertical accelerations on a shallow deadrise craft, to levels even below that of a similar craft with a steeper deadrise.

4.3 SHOCK MITIGATION

The reduction in vertical acceleration peak levels allows for synergisation with the use of suspension seating. When subjected to large accelerations which exceed the maximum shock displacement allowed by the seat's suspension mechanism, some designs of seat can bottom-out. This sudden stop amplifies rather than dampens the levels of shock to which the occupant is exposed [17]; using ballast to reduce the peak accelerations experienced by the craft as a whole [13], assists the suspension seat in operating as intended.

It is significant that the accelerations are reduced at the bow, because subject to the deck layout and passenger positioning, impact related injuries can be more common at the bow of the boat [7].

4.4 FURTHER STUDY

Further study is required to address a range of questions that were not covered in this simulation, and to identify potential routes for optimisation. The added mass of 12% was selected based on previous studies, but it warrants further exploration to weigh the balance between having larger amounts of ballast in order to achieve greater effects, against the amount of below-deck space taken up by the tank. The potential reduction in speed from greater amounts of ballast should also be explored.

In some cases, it may not be practical or possible to position the tank directly in the bow of the boat, so the effect of ballast positioned closer to the LCG should be examined.

Studies have shown that shock from sudden roll motions as the result of vertical impact can be a source of pain and discomfort [18], whilst the increase of mass has been proven to dampen such roll motions as well as the vertical pitching motions [19]. Further study could examine whether these roll movements could also be minimised through the use of dynamic ballast, as indicated by the results of Townsend et al [13], showing reductions in acceleration across each axis.

5. CONCLUSIONS

It must be stressed that the dynamic model requires validation using real world sea-trial testing. Given that this simulation model is a simplification of a complex issue, results should be taken with a degree of caution. However, these initial findings offer at least an indication of the benefits of dynamic ballast on high-speed craft for trim control and shock mitigation, as well as the potential for new avenues of design considerations.

The results of the above simulations, coupled with previous findings on dynamic ballast by Townsend et al [13] and Garme et al [14], offer interesting prospects for HSC designs. There is a clear reduction in vertical acceleration levels through the use of dynamic ballast, which remains effective across a variety of craft designs and operating conditions. Dynamic ballast could therefore potentially help designers to bridge the gap between craft optimised for high-speed calm sea transit, and those designed for rough sea operations.

6. ACKNOWLEDGEMENTS

The authors wish to thank Dr Trevor Dobbins, for his assistance in providing comments on a draft of this paper.

7. REFERENCES

1. ENSIGN, W., HODGDON, J., PRUSAZYK, W., SHAPIRO, D., LIPTON, M., 'A survey of self-reported injuries among special boat operators', *Naval Health Research Centre*, 2000

2. RILEY, M., HAUPT, K., JACOBSON, D., 'A deterministic approach for characterizing wave impact response motions of a high-speed planing hull', *Naval Surface Warfare Centre Carderock Division*, 2012

3. MYERS, S., DOBBINS, T., KING, S., HALL, B., AYLING, R., HOLMES, S., GUNSTON, T., DYSON, R., 'Physiological consequences of military high-speed boat transits', *European Journal of Applied Physiology*, 2011.

4. CAMILLERI, J., TAUNTON, D., TEMAREL, P., 'Slamming impact loads on high-speed craft sections using two-dimensional modelling', *Analysis and Design of Marine Structures*, 2015

5. DOBBINS, T., HILL, J., MYERS, S., 'Fatigue in military operations; high speed craft repeated shock and other factors', *Conference on human responses to vibration*, 2016

6. HENGST, D., CARPENTER, R., 'Impact induced injuries onboard high-speed boats', *HFM Experimental Task ET-183*, 2019

7. SELLIER, A., AL FALASI, M., JOUBERT, C., DESSE, N., BEUCLER, N., RENARD, A., BERNARD, C., ALLANIC, L., DAGAIN, A., 'A summer wave of vertical fractures: the "deck-slap injury', *Acta Neurochirurgica Spine Trauma Collection*, 2018

8. DOBBINS, T., MYERS, S., STARK, J., MANTZOURIS, G., 'Modelling human performance in maritime interdiction operations', *Human Modelling for Military Application*, 2010.

9. MYERS, S., DOBBINS, T., KING, S., HALL, B., HOLMES, S., GUNSTON, T., DYSON, R., 'Effectiveness of suspension seats in maintaining performance following military high-speed boat transits', *Human Factors: Journal of Human Factors and Ergonomics Society*, 2012

10. HAYNES, J., 'Developing a holistic strategy for shock mitigation on fast boats', *Shock Mitigation Next Gen Workshop*, 2016

11. SEBASTIANI, L., BRUZZONE, D., GUALENI, P., RAMBALDI, G., RUSCELLI, D., VIVIANI, M., 'A practical method for the prediction of planing craft motions in regular and irregular waves', *International Conference on Offshore Mechanics and Arctic Engineering*, 2008

12. GHADIMI, P., NOWRUZI, H., DASHTIMANESH, A., 'Parametric study of the effects of trim tabs on running trim and resistance of planing hulls', *Advanced Shipping and Ocean Engineering*, 2014

13. TOWNSEND, N., WILSON, P., AUSTEN, S., 'What influences rigid inflatable boat motions?', *Journal of Engineering for the Maritime Environment*, 2008

14. GARME, K., ROSEN, A., STENIUS, I., KUTTENKEULER, J., 'Rough water performance of lightweight high-speed craft', *Journal of Engineering for the Maritime Environment*, 2012

15. GRIMSLEY, J., 'Methodology to quantify vertical accelerations of planing craft in irregular waves', *Old Dominion University,* 2010

16. SAVITSKY, D., BROWN, P., 'Procedures for hydrodynamic evaluation of planing hulls in smooth and rough water', 1976

17. OLAUSSON, K., 'Vibration mitigation for high-speed craft', *Stockholm Royal Institute of Technology*, 2012

18. ALLEN, D., TAUNTON, D., ALLEN, R., 'A study of shock impacts and vibration dose values onboard high-speed marine craft',

19. TAVAKOLI, S., GHADIMI, P., DASHTIMANESH, A., SAHOO, P., 'Determination of hydrodynamic coefficients in roll motion of high-speed planing hulls', *International Conference on Fast Sea Transportation,* 2015

8. AUTHOR BIOGRAPHY

Joseph Salmon is Managing Director of Salmon Power Sports, specialists in dynamic ballast systems for high-speed craft. Salmon Power Sports have been designing and manufacturing dynamic ballast systems for racing, leisure, rescue, and professional craft for three decades.

Ben Landgren is a chartered engineer in the UK (RINA), and holds a Master's of Science degree in Marine Engineering from Newcastle University. His expertise lies in utilizing dynamic simulation models to design and evaluate marine engineering systems. Mr. Landgren has been working with companies of all sizes to help them understand complex engineering systems with the aid of dynamic simulation models.

PARAMETRIC OPTIMISATION OF A HYDROFOIL FOR A 24M HIGH-SPEED CATAMARAN FOR PASSENGERS TRANSPORTATION

M Minasi and **M Bovio**, Leading Edge Marine Engineering, Italy

SUMMARY

The subject of this study is the preliminary design of a hydrofoil-supported catamaran, aimed at improving the environmental sustainability of this means of transportation. The work is built around the study case of a 24m passenger ferry operating on the Cannes-Nice-Monaco-Ventimiglia route. Following the definition of an operating profile, a comparative analysis of typical catamaran hull shapes has been carried out using both analytical tools and higher-order CFD RANSE simulations to compute overall resistance values. For the scope, a brief Verification and Validation exercise is included as well. To evaluate the benefit of adding a hydrofoil for resistance reduction, an initial investigation of the 2D profile characteristics is presented, followed by further optimisation for higher speed application. In the present study, the profile shape optimisation is carried out using the PARSEC parameterization method, by coupling the panel method based XFOIL flow solver and DAKOTA software for the optimisation routine. A conclusive set of comprehensive 3D RANSE analyses of the complete system of hull and foil is then performed to validate the optimisation outcome.

NOMENCLATURE

υ	Kinematic viscosity (N s m^{-2})
ρ	Density of water (kg m^{-3})
P	Pressure (N m^{-2})
Rt	Still water overall resistance (N)
Rv	Viscous resistance (N)
Fr	Froude Number (-)
$Fr\nabla$	Volumetric Froude Number (m)
LCG	Longitudinal centre of gravity (m)
LCP	Longitudinal centre of pressure of the hydrofoil (m)
LOA	Length over all (m)
Lpp	Length between perpendiculars (m)
FP	Forward perpendicular (m)
FA	Aft perpendicular (m)
h	Hydrofoil immersion (m)
c	Chord Length (m)
Cl	Lift coefficient (-)
Cd	Drag coefficient (-)
Cp	Pressure coefficient (-)
τ	Trim angle (°)
δ	Flow direction with respect to the global reference system (°)
α	Hydrofoil angle of attack (°)
λ	Hydrofoil installation angle (°)

1. INTRODUCTION

The technology under examination in this paper is based on the concept of installing one or more hydrofoils, in the tunnel formed by the two hulls or below the keel line, in order to partially support the boat, to achieve an improvement in efficiency, and thus the reduction of the boat's advance resistance to motion and the reduction of fuel consumption and emissions, at high speeds. At low speeds, the presence of hydrofoils can lead to an increase in advance resistance compared to conventional boats, because the gain due to the hydrodynamic support generated from the profiles, which leads to a reduction in the boat's submerged volume, still does not compensate for the increase in resistance due to the presence of additional submerged bodies (i.e., hydrofoils).

First known systematic investigations were made by Soviet researchers in 1972, on asymmetrical semi-planning and planning hulls, fitted with a number of hydrofoils varying between 1 and 3. Poor information is provided in the literature on this subject, but it has been proven that the installation of two or three hydrofoils, reduces the advance resistance of the boat, for $Fr\nabla \geq 1.2$ [1].

Further investigations come from the development of the HySuCat platform, (i.e., Hydrofoil Supported Catamaran), led by Professor K.G. Hoppe at the University of Stellenbosch, South Africa, which resulted in the launch of the first prototype of submerged hydrofoil assisted catamaran in 1982, after which multiple boats were built using this technology [1]. In time, numerous other studies have been carried out for different hydrofoil configurations.

Main applications of this construction solution mainly involve fast passenger boats, but further examples can be found in high-speed pleasure craft and patrol boats, in the military.

2. MARKET ANALYSIS

The general dimensions and characteristics of the vessel were defined from the trend lines studied in the market analysis developed on 74 multihull vessels. The boats studied belong to the categories: catamaran, proa, trimaran, swath and hydrofoil supported catamaran as reported in Figure 1. The purposes of the analyzed boats are different, as shown in the Figure 2, of which mainly passenger boats were studied.

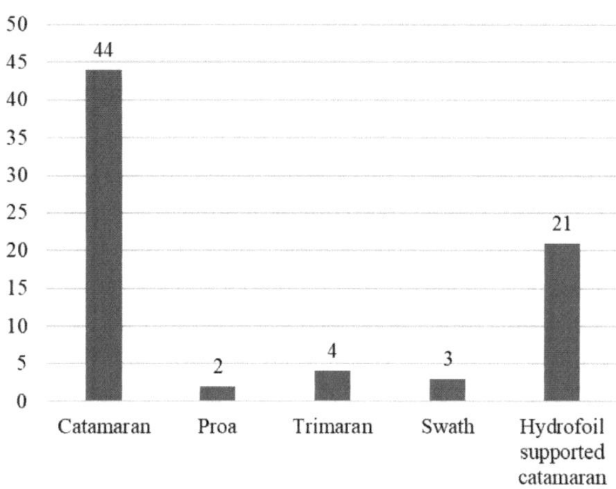

Figure 1: Multihull vessels categories studied

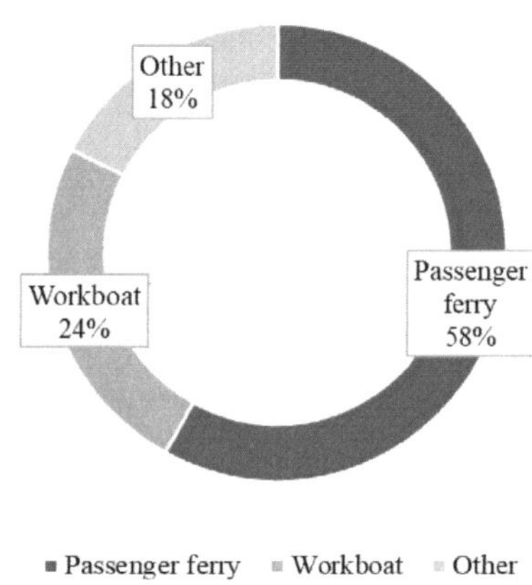

Figure 2: Multihull vessels functions studied

3. CASE STUDY

The case study is a passenger ferry catamaran HSC24, operating on the three routes: Monaco-Ventimiglia, Monaco-Nice and Monaco-Cannes. This territory is nowadays characterized by an increase in its population and by a strong development of its economy, hence leading to an increase of commuters, particularly to the Principality of Monaco. This is due to Monaco's economic vibrant activity, and also to the prohibitive pricing and scarcity of housing [2].

Operational profiles were assumed based on the data obtained from the market analysis and by comparing riding time by car and train with those of the HSC24 ferry. It was shown that the riding time of HSC24 is competitive with those from others means of transportation, Figure 3, and in particular they are lower in the case of the Ventimiglia-Monaco route. Moreover, it was shown that the impact of the fuel cost on the ferry ticket cost is similar to the fuel cost of car transportation, Figure 4.

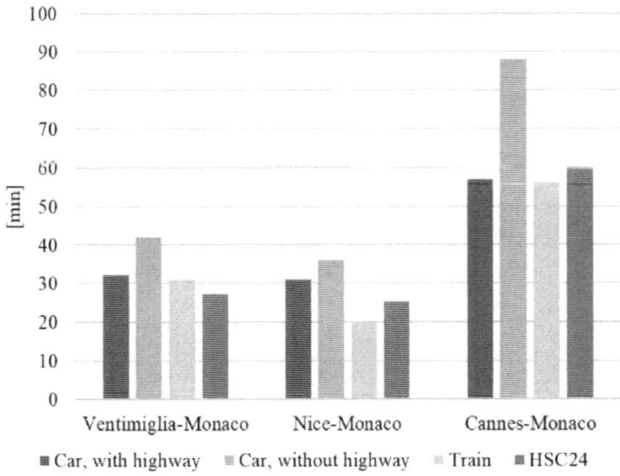

Figure 3: Riding time comparison

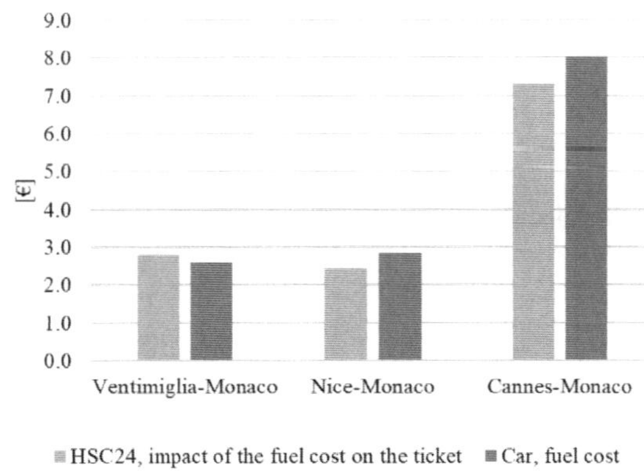

Figure 4: Impact of fuel cost comparison

4. NUMERICAL METHOD

4.1 CFD SOFTWARE

The resolution of Computational Fluid Dynamics (CFD) problems involves three main steps:

- spatial discretization of the flow domain,
- flow computation,
- visualization of the results.

To perform these steps, three dedicated software have been used. The first software (developed by Cadence), HEXPRESS™, is an automated all-hexahedral unstructured grid generation system. The second software (developed by the CNRS and the Ecole Centrale de Nantes), the flow solver ISIS-CFD, is a 3D unstructured flow solver able to simulate Euler or Navier-Stokes (laminar or turbulent) flows. The third software (developed by Cadence), CFView™, is a highly interactive Computational Field Visualization system. These three software have been integrated in a unique Graphical User Interface (GUI), called FINE™/Marine, allowing the solution of complete simulations of 3D internal and external flows from the grid generation to visualization [3].

4.2 NUMERICAL MODEL OF THE HULL

4.2 (a) Geometry and Reference data

The hull geometry used to carry out the Validation procedure is the Delft372 catamaran [4]. The comparison was made based on the experimental results of the still water resistance and wave profiles measurements done by A.P. van't Veer [4]. Wave profiles generated by the hull for the speeds highlighted in Figure 5, are compared. Longitudinal wave profile measurements are taken at 275 mm outwards from the demi-hull keel line [4].

4.2 (b) Computational domain and Boundary conditions

The domain size is defined by the size of the vessel with the following proportions: X_{max} = LOA *1.5, Z_{min} = LOA*1.5, X_{min} = LOA*3, Z_{max} = LOA, Y_{max} = LOA*2, Y_{min} = 0.
The boundary conditions imposed are:

- *Solid* condition, for each surface that is part of the hull. On the hull and at transom, a *wall function* approach is applied, while on the deck the *slip* condition is imposed. The *slip* type of boundary condition corresponds to zero tangential stresses on the wall, so it neglects the effects of turbulence.
- *Symmetry plane* condition, for the domain surface dividing the real model in half. This condition assumes that the geometry and the flow field are symmetrical with respect to this surface.
- *External*, for the remaining domain surfaces. Specifically, the *far field* condition is set on the remaining vertical surfaces, where the flow velocity is that of the free stream velocity. For the horizontal surfaces, the condition applied is *updated hydrostatic pressure*, which calculates the pressure along the axis oriented in the direction of gravity acceleration and updates this value as the simulation evolves, if the free surface varies.

4.2 (c) Validation Results

The resistance curve in still water and the wave profiles for the highlighted velocities with a red circle in Figure 5 are compared respectively in Figure 5 and 6.

The correlation between numerical and experimental results is well represented, particularly up to Froude numbers about 0.5. For higher velocities larger discrepancies are appearing, which could be attributed to experimental uncertainties due to the size of the towing tank facility, whose walls may have interfered with the pressure field generated by the model.

For values of Fr>0.5, in the semi-planing regime, the differences could be ascribable to blockage effects, and this can be confirmed by the fact that the experimental results have larger values of trim angle associated with a larger sinkage: at this speed regime, this is an interesting indication of a general lowering of the free surface height around the vessel, which is typical of blockage effects.

At the maximum velocity instead, which is 4.06 m/s, the differences between the results may be due to the fact that we are in the subcritical region identified by the Froude number value calculated over the tank depth $Fr(h)=V/\sqrt{gh}>0.7$, where bottom effects are not negligible [9].

To validate these hypotheses, additional simulations should be carried out, in which the actual walls of the towing tank are also reproduced.

The wave profiles are also well represented, and the biggest differences are for the two higher velocities where the numerically calculated wavelength is greater than the experimental one, therefore also the phases of the two profiles are slightly different.

In the light of above-mentioned observations, excluding the higher speed cases, the numerical model is deemed as validated against the experiments, considering satisfactory the correlation between the two sets of data.

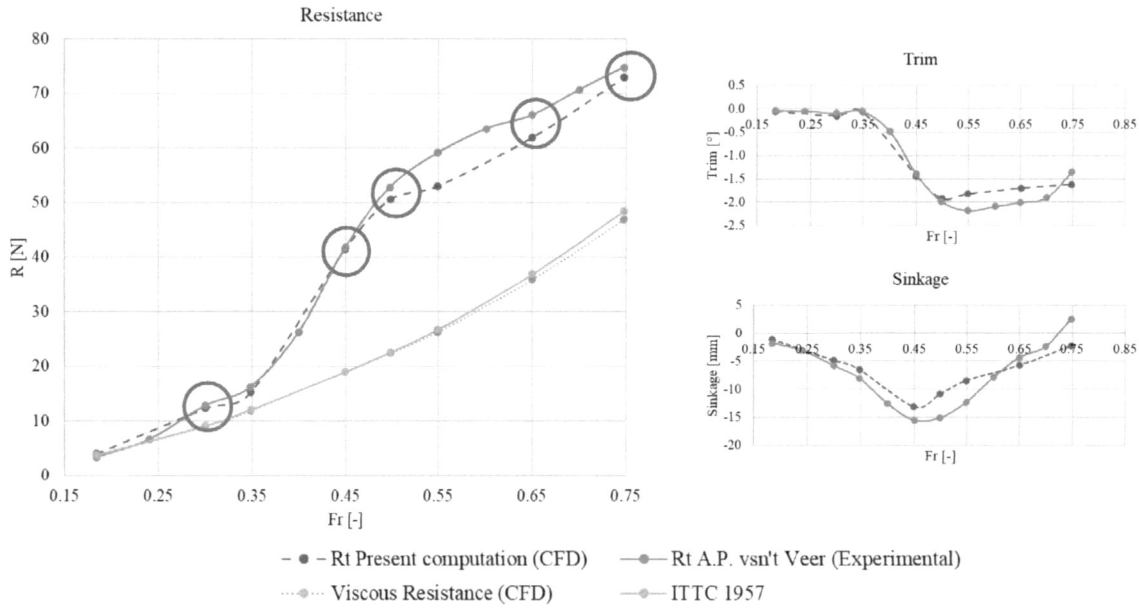

Figure 5: Still Water resistance curve, sinkage and trim comparison

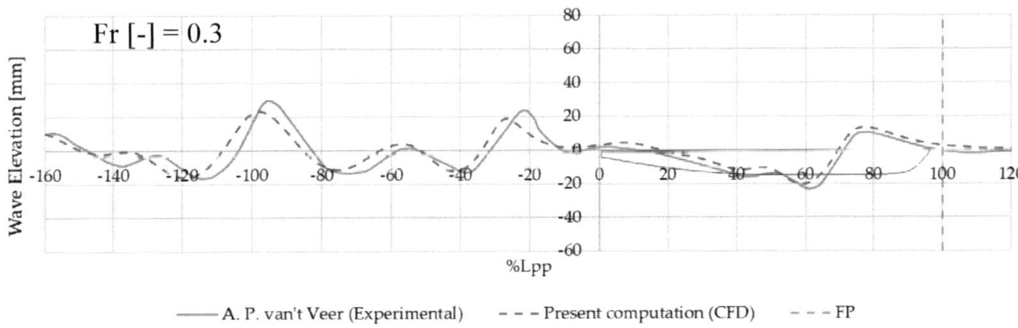
Fr [-] = 0.3
Wave Elevation [mm]
%Lpp
A. P. van't Veer (Experimental)
Present computation (CFD)
FP

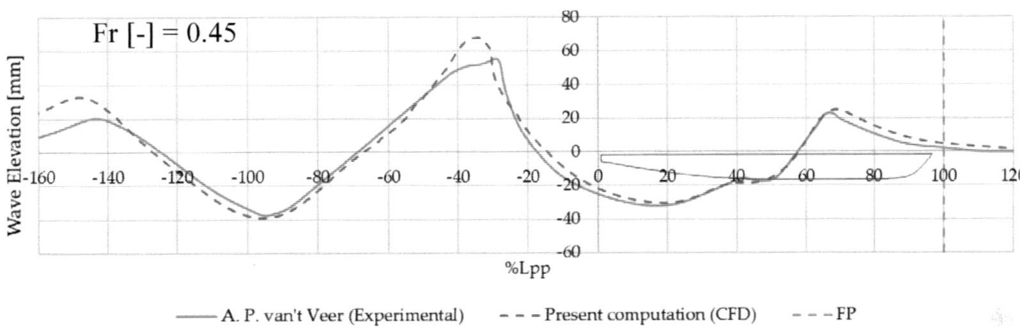
Fr [-] = 0.45
Wave Elevation [mm]
%Lpp
A. P. van't Veer (Experimental)
Present computation (CFD)
FP

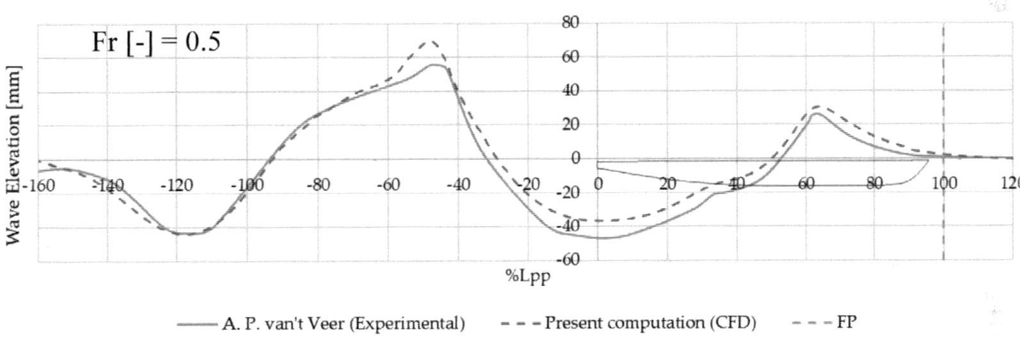
Fr [-] = 0.5
Wave Elevation [mm]
%Lpp
A. P. van't Veer (Experimental)
Present computation (CFD)
FP

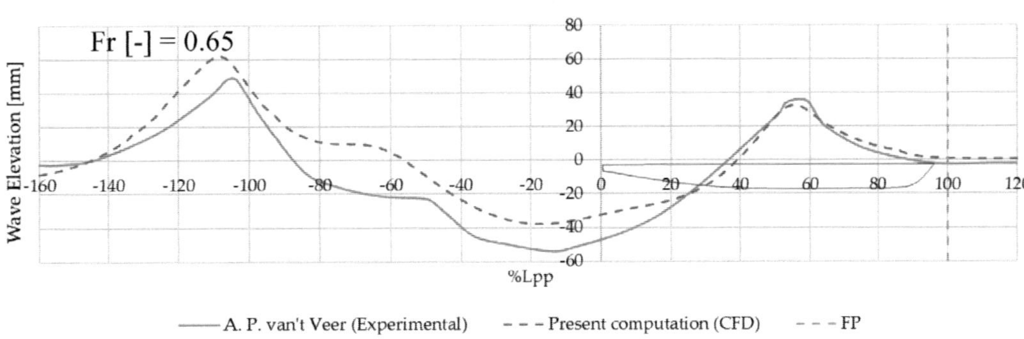
Fr [-] = 0.65
Wave Elevation [mm]
%Lpp
A. P. van't Veer (Experimental)
Present computation (CFD)
FP

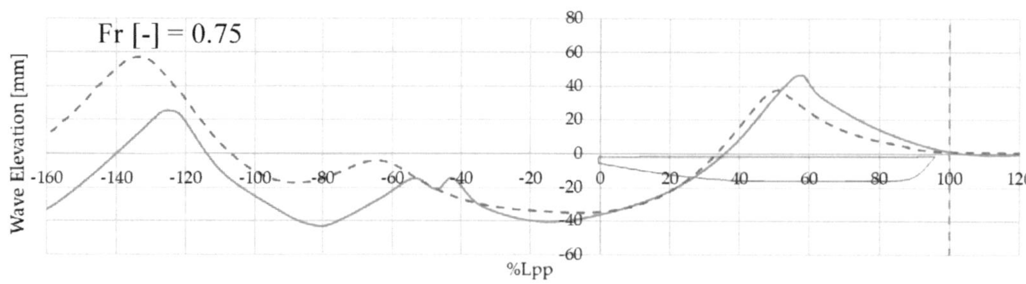

Figure 6: Wave profiles comparison with experimental results from the study of A.P. van't Veer

4.3 NUMERICAL MODEL OF THE HYDROFOIL

4.3 (a) Geometry and Reference data

Verification and Validation procedures were carried out comparing the numerical data with J. Duncan's experiment on the free-surface proximity effects on the efficiency of an immersed wing. [6]. The reference test case consists in the measurements of the wave profiles generated by a NACA0012 of chord equal to 0.203 m while moving parallel to the free surface at a velocity equal to 0.8 m/s, for several h/c values as shown in Table 1. The position of the interface is defined through the immersion depth h between the initial position of the interface and the position of the centerline at mid-chord c/2, in order to compare with the reference literature [6].

Since the values of the hydrodynamic coefficients were not reported from the experimental tests, the numerical results were compared with the outcome of another numerical study from J. Wackers et al. [7].

4.3 (b) ` Computational domain and Boundary conditions

It was chosen to assign the foil surfaces a degree of refinement so as to obtain at least 70 cells per chord length and a cell height value on the free surface of maximum 3 mm height, so as to achieve about 13 cells per wave height to grant for a fair resolution of the wave elevation.

Table 1. List of h/c studied, in order to compare the numerical results with the J. Duncan's experimental results

h [m]	h/c [-]
0.26	1.29
0.24	1.16
0.21	1.03
0.19	0.95
0.19	0.91
0.16	0.78

The computational domain is defined as 10 chord lengths in the upstream direction and 20 chord lengths downstream, the height of the domain was imposed to achieve a discrete number of initial squared cells such as to better control the grid dependency study. The value 0.875 chord lengths below the profile was set to reproduce the same conditions as in the experimental tests; in fact, since the profile is close to the bottom of the tank, the bottom effect is not negligible.
The boundary conditions imposed are:

- *External*, at the inlet and outlet, with the undisturbed constant velocity condition of 0.8 m/s, and *updated hydrostatic pressure*, for the upper surface of the domain.
- *Solid Slip* condition on the bottom is applied.

- *Solid wall function*, on the face and back of the profile the calculation of wall functions is assigned.

4.3 (c) Verification Results

The Verification procedure was carried out for the maximum value of h/c=1.29, because for lower immersion values, especially for h/c≤1, nonlinear phenomena, such as the generation of breaking waves, are encountered, thus introducing numerical noise in the procedure.

The Verification of the numerical model of the hydrofoil was carried out using the Richardson Extrapolation in combination with the Grid Convergency Index as reported in [5], obtaining asymptotic convergence. The numerical uncertainty for grid refinement used in the simulations was calculated to be 0.02 and corresponds to the *fine* mesh refinement. The characteristics of the *fine* mesh refinement was previously described in chapter 3.3 (b).

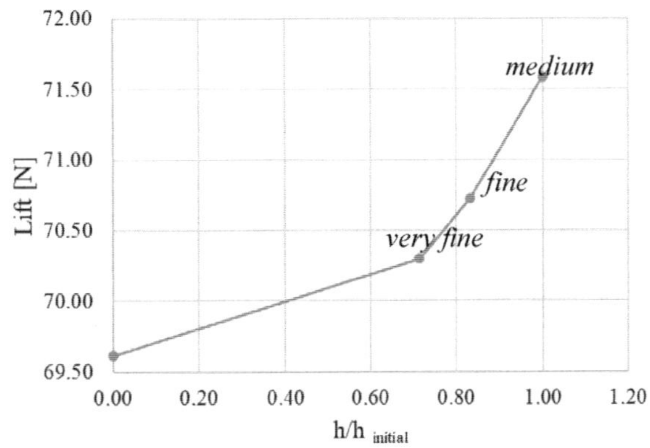

Figure 7: Asymptotic convergence obtained by Richardson Extrapolation procedure

4.3 (d) Validation Results

The wave profiles computed are compared with the experimental results, reported in Figure 8, and are qualitatively well represented. The largest differences occur for the two lower h/c values and are likely due to the phenomenon of breaking waves, which numerically implies that the result is time dependent, therefore not compatible with a steady state approach. Additionally, it has to be noted that even experimentally it is difficult to define the free surface in the presence of foam.

The hydrodynamic coefficients are compared with the numerical results calculated in the study by J. Wackers et al. [7], Figure 8, for immersion values studied by Duncan [6] and also for h/c=2. The trend of the hydrodynamic coefficients is in line with the numerical results by Wackers [7], there is a gap between the lift coefficient results, probably due to a different numerical setup and the use of the Adaptive Grid Refinement feature which was not applied in this study for an easier control of the grid refinement ratios. The influence of the free surface proximity results in a drastic drop in the immersed profile efficiency, due to the increase in resistance and the loss of lift. For h/c values below 0.8, the influence of the free surface has therefore considerable effects.

Given the well represented trend of the results, the numerical model is considered validated against the experimental case.

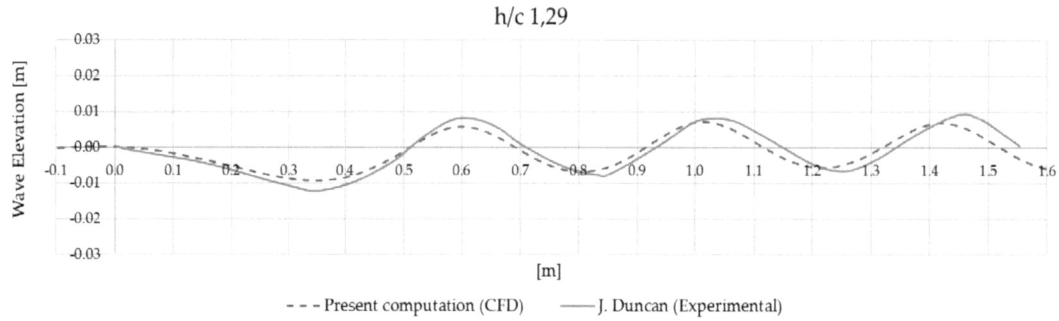
h/c 1,29
Wave Elevation [m]
[m]
- - - Present computation (CFD)
—— J. Duncan (Experimental)

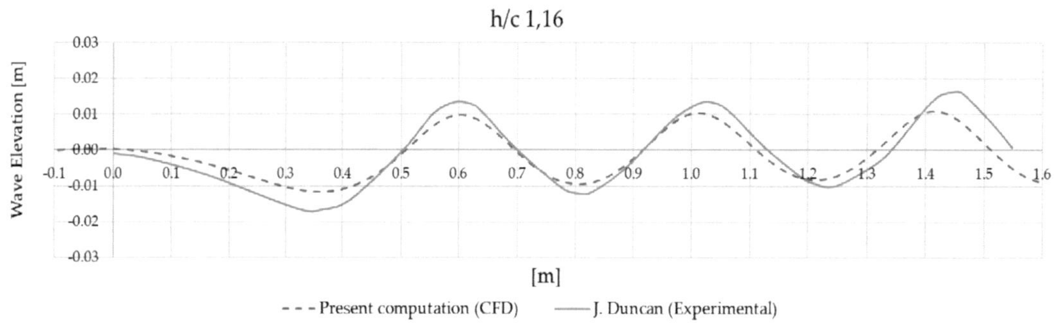
h/c 1,16
Wave Elevation [m]
[m]
- - - Present computation (CFD)
—— J. Duncan (Experimental)

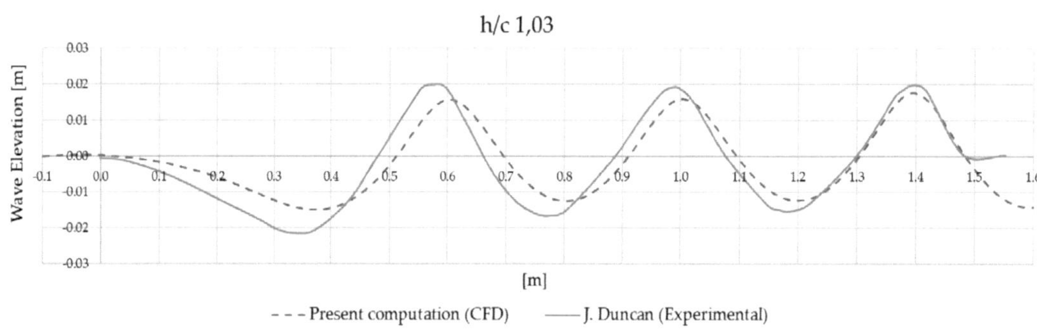
h/c 1,03
Wave Elevation [m]
[m]
- - - Present computation (CFD)
—— J. Duncan (Experimental)

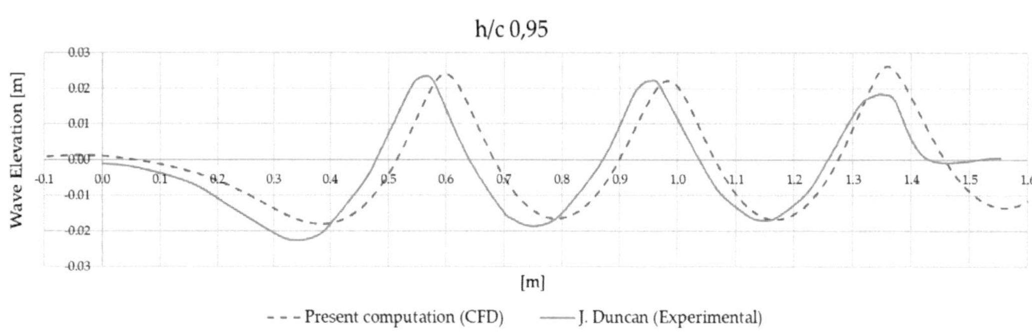
h/c 0,95
Wave Elevation [m]
[m]
- - - Present computation (CFD)
—— J. Duncan (Experimental)

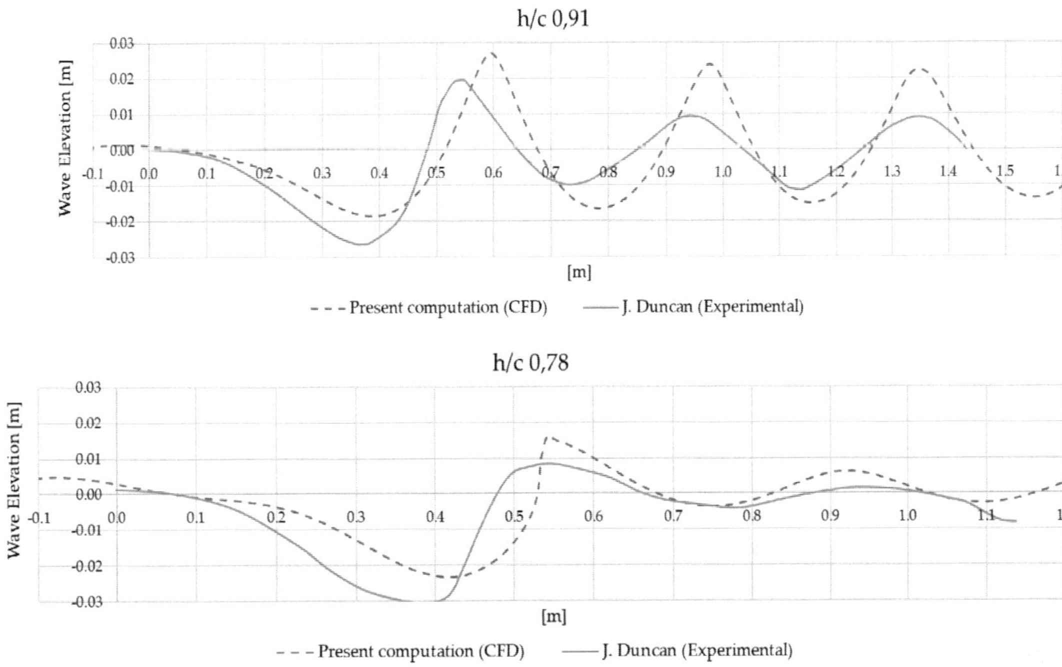

Figure 8: Wave profile computed compared to the experimental results from J. Duncan

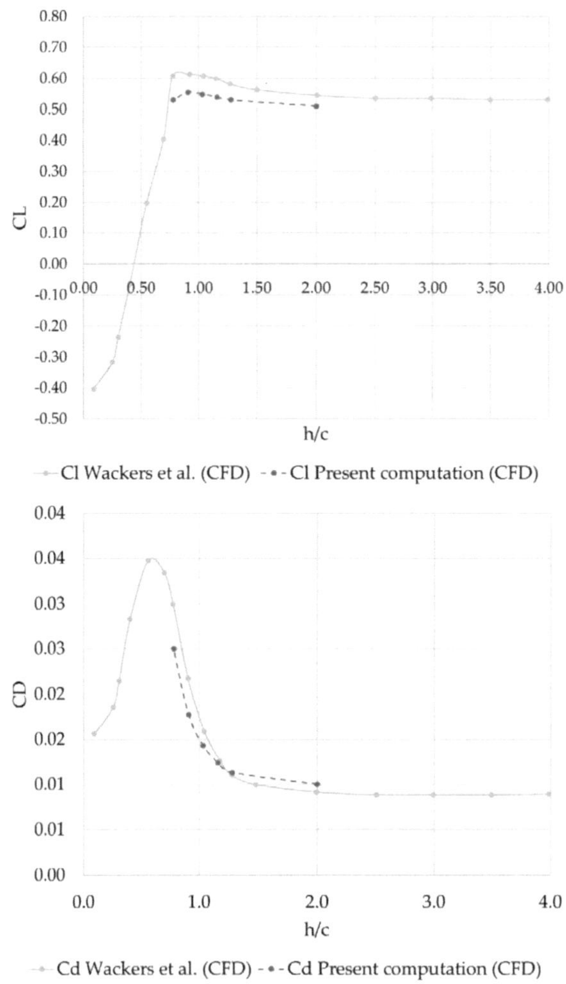

Figure 9: Computed hydrodynamic coefficients compared with the numerical results from Wackers et al.

5. HULL DESIGN

Two different hull shapes are then designed, one symmetrical to have a comparison with the literature [1], and one asymmetrical in line with other boats as found in the market. Following a performances analysis, it was chosen to proceed in the study with the asymmetric configuration, since on average the performance was deemed to be more satisfactory, Figure 10.

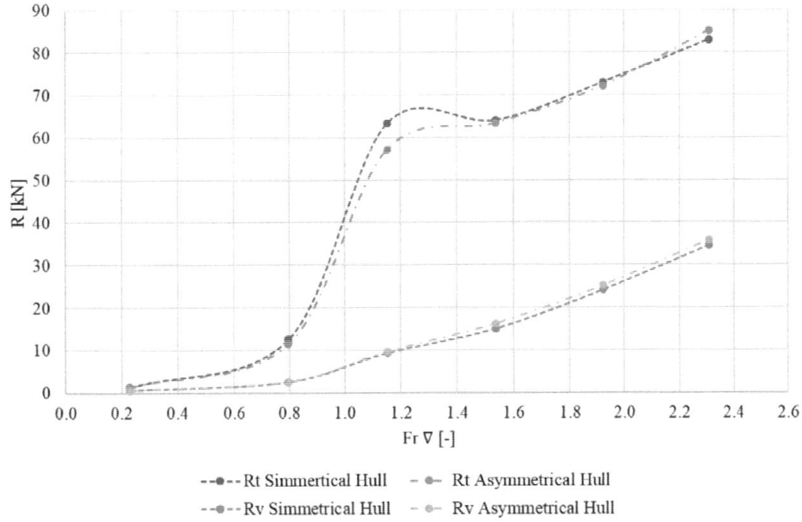

Figure 10: Still water resistance comparison between Symmetrical and Asymmetrical hull designs

6. HYDROFOIL DESIGN

The hydrofoil dimensions were chosen based on both structural and hydrodynamic considerations. The hydrodynamic analyses were performed with the opensource software XFOIL. A single-hydrofoil configuration is chosen in this study, whereby a single wing is installed in the tunnel enclosed between the two hulls. Four profiles were then analysed, (Eppler817, NACA65410, YS915, Speer H105) mainly chosen because they were originally designed to avoid cavitation at high speeds.

Having initially defined the dimensions in terms of span, as dictated by the distance between the two hulls, the chord dimensions were defined based on structural considerations together with the lift required to the wing. An internal support structure has been sketched, and for various types of sections and materials (rectangular hollow, rectangular solid, circular solid, each in steel and aluminium), it was verified that only in the case of a chord equal to 1.5 m, the maximum tension would be less than the yielding stress. The target load for which the criterion was verified, is equal to a force uniformly distributed over the span of the profile, taken equal to 40 % of the displacement, in line with data from literature [1].

Established the chord size, the performance of the different profiles at the design speed of 30 kn was assessed and results are shown in Figure 11.

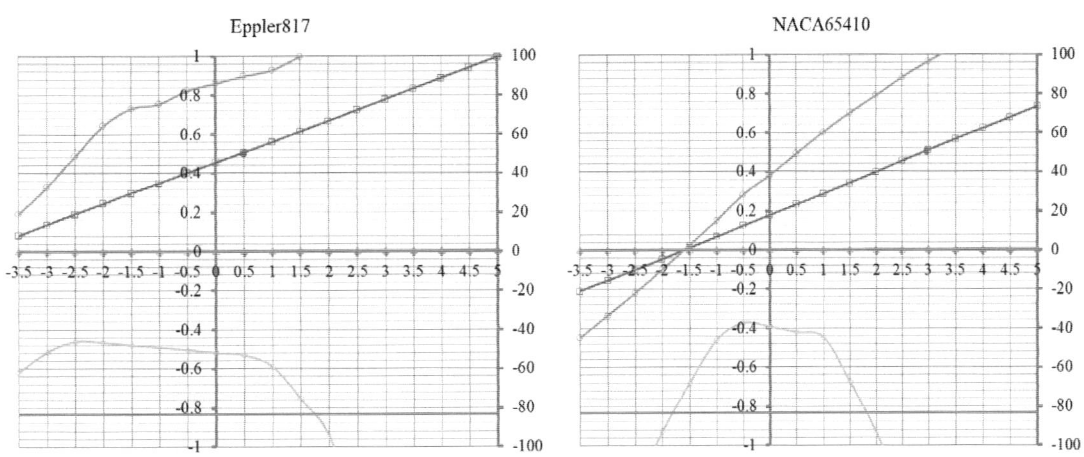

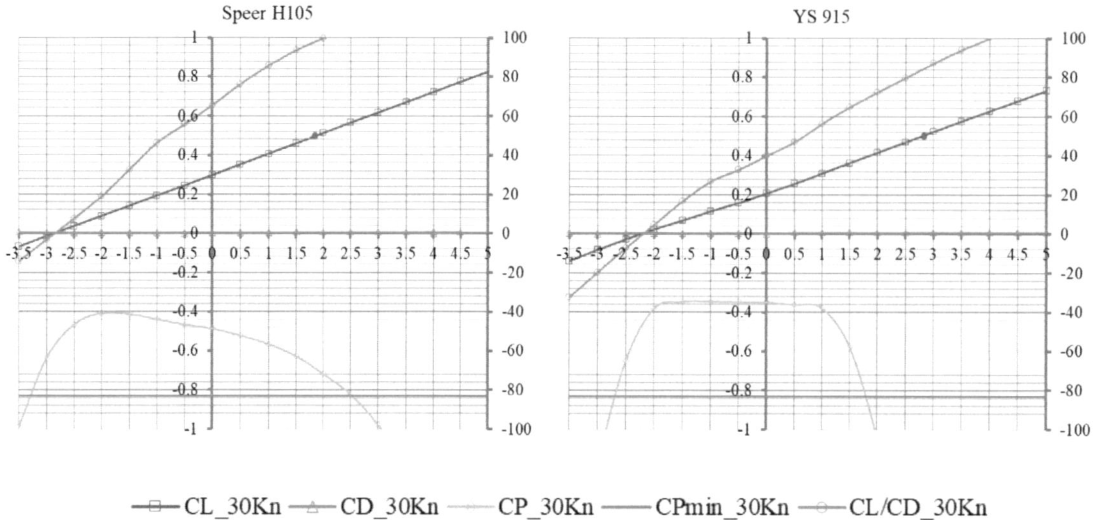

Figure 11: Performance analysis from the 2D profile study. The red dot represents the target Cl to achieve a Lift corresponding to the 40% of the total displacement of the vessel.

Excluding the NACA 65410 and YS915 profiles, since their cavitation buckets are too narrow for the intended use, the Eppler817 over the Speer105 profile was chosen to be installed on board, since it had more margin over the cavitation limit while being similar in terms of hydrodynamic efficiency.

7. HULL AND HYDROFOIL INTERACTION

The effects of the hydrofoil installation on the performance of the boat are studied using the single-hydrofoil configuration and initially considering the longitudinal position of the wing centre of pressure equal to the longitudinal position of the overall centre of gravity (LCP=LCG). Since the dynamic equilibrium assumed by the boat following the installation of the wing cannot be known a priori, a simplified numerical approach was initially carried out. A vertical force equal to 40% of the displacement, as defined previously, was then applied in order to better estimate to final dynamic equilibrium of the vessel under the wing influence and to define the incident angle of the flow lines relative to the leading edge of the wing to estimate what the actual operating condition of the hydrofoil will be. The results are reported in Table 2.

The hydrofoil is installed at a distance of 10 cm above the keel line to take advantage of the wall effect of the inboard side, also, to ease the typical drydocking operations. The influence of the free surface proximity on the performance of the hull must be defined since the hydrofoil immersion in operational mode is not known. Therefore, the performance of the complete hull+hydrofoil system was studied to correlate the three-dimensional performance with the estimated two-dimensional performance, so to define a correction factor to account for the influence of the free surface in the following phases of the study.

7.1 VARIATION OF ANGLE OF ATTACK

As reported in Table 2, several simulations were run to evaluate the effective angle of attack of the wing. In *simulation a*, having defined δ and τ from the numerical hydrofoil simulation, the hydrofoil is installed such as the angle of attack α is 0.5 with respect to the incoming flow, as required for the selected Eppler 817 profile from the 2D study to achieve the required lift of 40% of displacement.

From the results of *simulation a* is then measured the actual angle of attack α, which was found to be equal to 0.23°. Then *simulations b* and *c* were carried out by varying the angle of attack and setting it at 0.75° and 1.5°, and consequently varying the installation angle λ which is referred to the keel line. The angles used and related results are given in Table 2. The hydrofoil installation angle, as well as the vessel trim angle, both follow the right had rule convention, and is defined as shown in Figure 12.

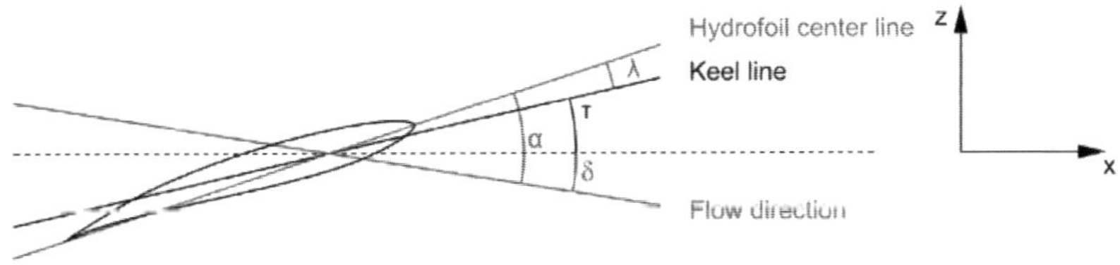

Figure 12: Example of angle definition

As the angle of attack increases, the drag reduction increases to a value of 9.55%, which corresponds to a generated lift equal to 19.02% of the displacement. The result, however, is still far from the reference value of 40% of displacement.

Most of the cause is probably ascribable to the proximity of the free surface given the h/c ratio is 0.56, when using the same reference system used by J. Duncan [2]. This value is lower than the critical value h/c = 0.8, below which the influence of the free surface results in the drastic drop in profile performance. Comparing the performances obtained from the 3D study with the performance achieved at the same angle of attack in the 2D study, the profile performance turns out to be about 60 % lower than in the two-dimensional case, for all the cases as investigated. However, this implies that even when working at angles of attack greater than those imposed by the cavitation limit defined in the two-dimensional study, the hydrofoil works under conditions far from cavitation risk.

The study was therefore completed using the configuration studied in *simulation 1c*, with an angle of attack of 1.5°, which is the most advantageous in terms of resistance reduction at the design speed. In Figure 13 the resistance curves are shown.

In the first part of the curve the resistance of the hull with the wing is greater, whereas an actual reduction in resistance can be appreciated for speed values greater than 20 kn.

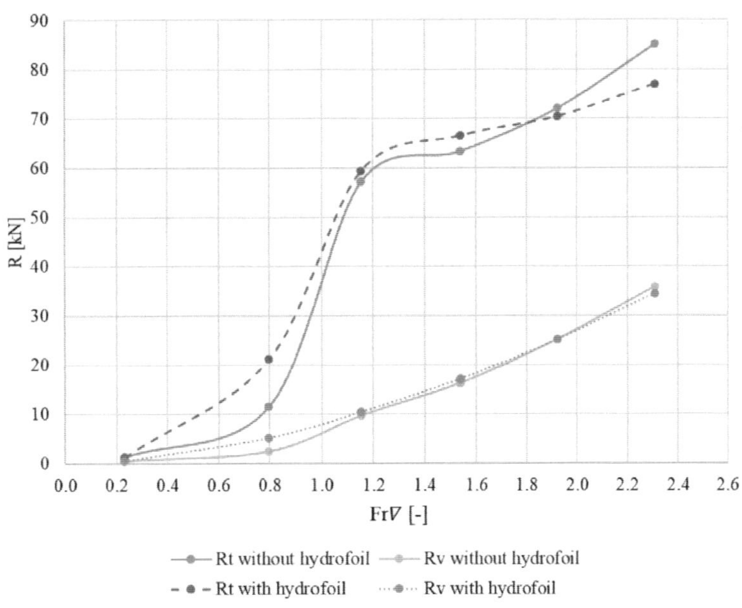

Figure 13: Still water resistance comparison with and without hydrofoil

7.2 VARIATION OF LCP

Following the first investigation on the angle of attack, a further assessment of the influence of the longitudinal centre of pressure, at the design speed, was carried out. Two conditions were tested by shifting the LCP of 2% of the LWL in both ahead and astern directions, from which it appears that moving forward results in an additional drag reduction and a lift equal to 23,99% of the displacement. The results are reported in Table 2. This configuration was chosen to study the consequences of installing the optimized profile.

Table 2. Results and data used for the hull-hydrofoil system simulations, varying the angle of attack and LCP position

	τ [°]	α [°]	λ [°]	Rt [kN]	%Rt reduction	%Δ reduction	Heave [m]
Simulation without hydrofoil	-1.51	[-]	[-]	85.15	[-]	[-]	0.07
Simulation with numerical hydrofoil	-1.33	[-]	[-]	22.92	[-]	[-]	0.46
Simulation a	-1.41	0.25	1.21	79.57	6.55	13.03	0.19
Simulation b	-1.43	0.75	0.69	78.31	8.03	15.42	0.21
Simulation c	-1.57	1.50	-0.06	77.01	9.55	19.02	0.25
Simulation d (LCPfwd)	-1.82	2.31	-0.07	75.87	10.90	20.44	0.28

8. 2D PROFILE OPTIMIZATION

To carry out the profile optimization, the Bezier-Parsec parameterization method was used to describe the profile geometry, and the Dakota software optimizer coupled with the XFOIL flow solver based on the panel method was used. By combining the parametric description of the profile shape, the multi-objective genetic algorithm and the flow solver, the new hydrofoil shapes are derived. This parameterization method uses Parsec parameters to define four separate Bezier curves. These four curves define the leading edge and trailing edge of the camber line and of the thickness distribution. In this study the BP 3434 parametrization was used; in BP 3434 parameterization both the leading edge curves have polynomials of degree three while both the trailing edge curves have the polynomials of degree four. The BP 3434 parameters are shown in Figure 14.

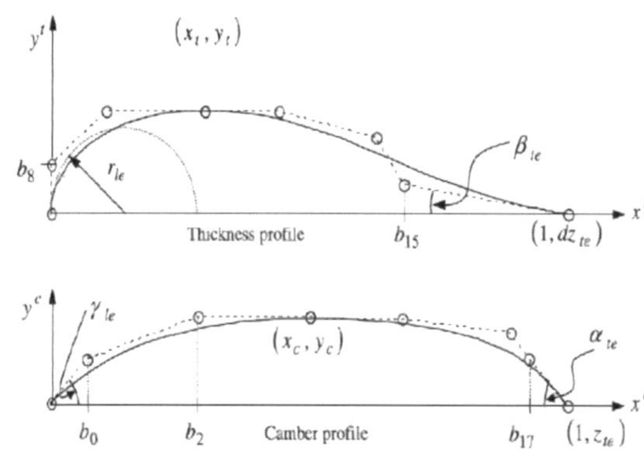

Figure 14: BP 3434 airfoil geometry and Bezier parameters and control points

A total of 11 out of 15 variables were allowed to vary in order to meet the structural limits imposed and giving sufficient morphing capabilities for the optimizer to explore the design of experiment space.

Consequently, the Dakota optimizer defines an initial population consisting of 150 profiles, whose parameters are randomly defined within the imposed limits. The optimization process leads to the typical Pareto fronts in accordance with the defined objectives. The objectives of the optimization are set as to minimize the angle of attack and related drag coefficient for an imposed lift coefficient. A required lift coefficient increased by 50, 60, 70 % from the value defined in the two-dimensional study was calculated to account for the lift losses due to free surface proximity as previously detected.

The resulting Design of Experiments and related Pareto fronts for the three different cases are reported in Figure 15. According to the optimization routine, the best profile sections approximatively coincide for the three cases. The characteristics of the optimized profile compared with the reference one is shown in Figure 16.

From δ and τ obtained in *simulation d*, the installation angle λ was defined to reach the required angles of attack in the three different cases, to obtain the required lift coefficient increased by 50, 60, 70 %.

The 2D performance analysis of the optimized profile is shown in Figure 17. Angles of attack greater than 1.5° fall outside the cavitation limit imposed by the two-dimensional study, however, given the influence of the free surface it is expected that the hydrofoil will work well below the limit imposed by cavitation. As a result of the profile optimization, an additional gain in drag reduction of about 1.5% was achieved.

It can be observed that as the angle of attack of the optimized profile increases, it results into a drag reduction of up to 12.32 % when compared with the configuration without the foil. Together with a lift increase, also the drag generated by the foil increases, but in the total drag balance there is still an advantage due to the reduction in hull drag associated with its emersion. The results are shown in Table 3.

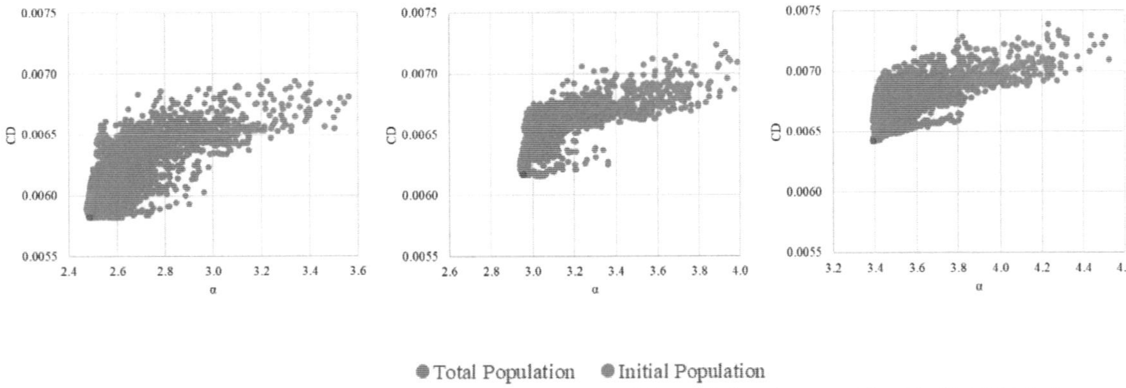

Figure 15: Pareto fronts of the optimisation for a target lift coefficient increased by 50, 60, 70% (left to right). In the graphs is also highlighted in orange the initial population composed by 150 randomly generated profiles for each of the cases. The red dot indicates the best profile section which meets the requirements of the objective functions.

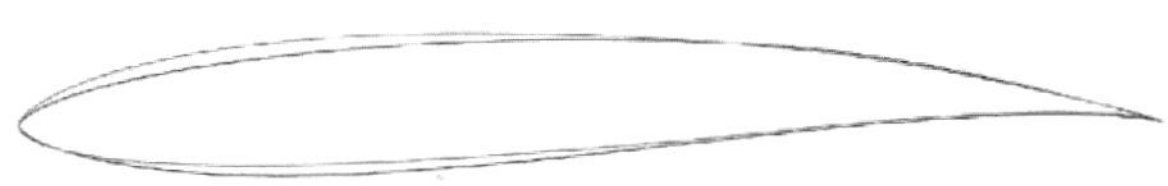

Figure 16: Optimized profile (green) and Eppler817 (black) comparison

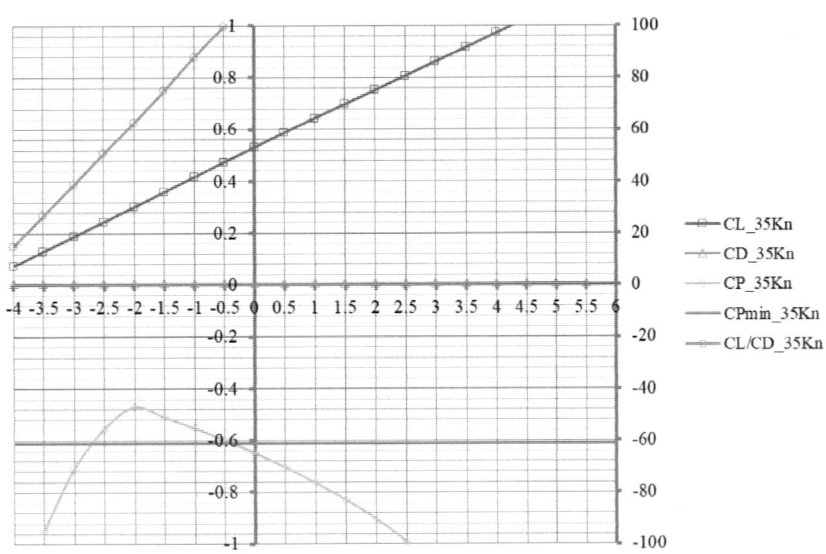

Figure 17: Two-dimensional performance analysis of the optimized profile

Table 3: Results and data used for the hull+optimized hydrofoil simulations

	τ [°]	α [°]	λ [°]	Rt [kN]	%Rt reduction	%Δ reduction	Heave [m]
Simulation CL50	-1.90	2.49	-0.25	75.63	11.18	21.30	0.29
Simulation CL60	-2.02	2.96	-0.72	74.88	12.06	24.06	0.32
Simulation CL70	-2.09	3.39	-1.16	74.66	12.32	25.59	0.34

9. CONCLUSIONS AND FURTHER DEVELOPMENTS

As a result of this study, a 12.32% reduction in advance resistance was achieved, at the design speed of 30 kn. Further investigation may be to extend the optimization study to additional speed values, with the aim of reducing resistance even at lower speeds.

With the profile optimization procedure, it was possible to achieve an additional gain in terms of drag reduction of 1.5%. This procedure is therefore effective, and the use of the XFOIL flow solver coupled with the Dakota optimizer can be applied conveniently, especially at a preliminary design stage, from the point of view of computational time, when compared with the use of CFD RANSE simulations.

It was shown that the influence of the free surface prevents the target lift value from being reached. By increasing the hydrofoil angle of attack, and consequently increasing vessel heave, the back surface of the foil gets closer to the free surface resulting in a design limitation. It is therefore possible to arrive at an operating condition beyond which increasing the angle of attack can only constitute an additional increase in drag, not becoming a gain.

To overcome this problem, it is possible to consider a different hydrofoil configuration, such as moving the wing below the keel line. This results in an increase in the h/c ratio and consequently a reduction of the detrimental effects associated with the free surface proximity. However, in this case it must be considered that the presence of the three-dimensional vortex effects at the wing tip will be more influential.

10. REFERENCES

1. MIGEOTTE, G., 'Design and Optimization of Hydrofoil-Assisted Catamarans', 2002.
2. LGV PROVENCE ALPES COTE D'AZUR, 'Collegamento ferroviario Nizza Italia', 2012.
3. CADENCE, 'FINE Marine 11.1 User Guide'.
4. VAN'T VEER, A P., 'Experimental results of motions, hydrodynamic coefficients and wave loads of the 372 catamaran model', 1998.
5. CELIK, I B., et al., 'Procedure for Estimation and Reporting of Uncertainty Due to Discretization', 1993.
6. DUNCAN, J., 'The breaking and non breaking wave resistance of a two dimensional hydrofoil', 1983.
7. WACKERS, J., et al., 'Free surface effects on two dimensional hydrofoils by RANS-VOF simulations', 2021.
8. DERKSEN, R W., et al., 'Optimum aerofoil parameterization for aerodynamic design', 2009.
9. LARSSON, L., and RAVEN, H C.,'The Principles of Naval Architecture Series', J. Randolph Paulling Editor, 2010.

11. AUTHORS BIOGRAPHY

Maddalena Minasi is an intern at Leading Edge Marine Engineering and has recently graduated with a MSc in Naval Architecture. Part of the reported study was the subject of the master's thesis developed during the internship.

Marco Bovio holds the position of Naval Architect & CFD specialist at LeadingEDGE Marine Engineering consultancy firm. The company provides for cutting-edge solutions to the maritime businesses with creativity and efficiency. Main focus is on numerical simulations (Computational Fluid Dynamics and Finite Element Analysis) to support the designing and troubleshooting processes typical of the shipbuilding industry, both for commercial and leisure projects. Alongside, effective Project Management is achieved by integrating profound knowledge in the Naval Architecture and Marine Engineering fields.

A HOLISTIC CONCEPT DEVELOPMENT OF A LUXURIOUS AND SUSTAINABLE CRUISE VESSEL

M M de Monchy, Design studio IJgensinnig, Rotterdam the Netherlands and **F Feder**, FF arch architecture, Rotterdam the Netherlands

SUMMARY

This paper will show the holistic concept development of a comfortable cruise ship with the smallest footprint possible.

Based on the environmental awareness of the early cruise adaptors a new cruise market potential is created.
By analysing 100 cruise ships and 50 yachts, the most appropriate comfort/price ratio is determined. This comfort/price ratio and the environmental awareness of the early cruise adopters are taken into account, when creating the highest possible level of comfort on board.

The combination of these two aspects results in a new type of cruise ship that meets the requirements of the early adopters of exclusive cruising. Proposals for alternative materials and systems on board are based on the reduction of environmental impact as often used in the building industry.

Based on the market research and the elaborate experience of both authors in the field of sustainability, as well in the architectural and shipbuilding industry, a new type of cruise vessel is developed.

NOMENCLATURE

$L_{.o.a}$	Length over all (m)
V_{cruise}	Cruise speed (m/s)
Displacement	ton
Cost price per person per day	€/p.p.d.
Square meters per passenger	m^2/p
The comfort parameter	€/ m^2 p.p.d

1. INTRODUCTION

Over the past few decades, cruising has grown in popularity as ships have become larger to accommodate more passengers. As prices dropped, cruising became accessible to people with average incomes. Most of the cruise vessels are designed for the mass tourism and operate as efficiently as possible.
On the other hand the positive image of the cruising industry is declining due to its impact on the environment. The growing awareness of our planet with her sources and limits forces the industry to change.

To keep the early cruisers with above average financial resources, expecting luxury and ecological awareness as clients, a new type of cruise vessel is needed to be developed. The growth of this group has created a market potential for a new generation of cruise ships. In this paper the concept development of the new generation cruise vessel will be shown.

The holistic design process of these ships is based on extensive market analysis [1], observations, interviews of the cruise passengers and the extensive experience of the authors in the field of sustainable architecture.
Comfort and the environmental aspects of the developed concept form the starting of the design process, that will be discussed in the following chapters.

2. COMFORT

2.1 DEFINING COMFORT

Before starting to design a comfortable cruise ship, comfort and luxury was defined first. After analysing the cruise market and the private yacht charter market, luxury and comfort were described with parameters.

Like mentioned earlier most of the cruise vessels are focussed on mass tourism. These vessels are built as cheap and as efficient as possible to make it accessible for a lots of people. The so called early adaptors of cruising are used to more space and less people on board. They miss the luxurious cruises from decades ago. Chartering a private yacht is for them too expensive, renting just one cabin on board of a private yacht will be impossible.
We discovered that luxury means having more space per person and less persons onboard of a cruise vessel. The less passengers on board of the ship, the higher the cost prices of the cruises will be. So we could assume that comfort can be defined by luxury (chapter 2.1 b) on board of the cruise vessels and the cost price of these cruises (chapter 2.1 a).

2.1 (a) Determining Cost price

Analysing different cost prices of the cruises from different providers, different price classes were noticed. To make it easy to compare different cruise prices with each other, the cost price is determined on prices a day per person on board (Cost price per person per day €/p.p.d.). Besides creating this parameter for cruise trips, this parameter is also created for the private charters. In this case a comparison can easily be made. For having a global inside more than 100 cruise trips were compared. Due to the exclusivity smaller market share, just 50 yacht charters are taken into account.

For each cruise ship 3 different price categories were investigated; 2 person cabin inside, 2 person cabin outside and the so called penthouses. For these 3 different categories the price per person a day was calculated.
After analysing the data and the €/p.p.d., we conclude that the private charter is to expensive.

2.1 (b) Determining Luxury

After interviewing the so called early cruisers, two aspects of luxury were claimed; the amount of crew on board and the available space on board. In general the more people on board the more space and crew available on board. According to the interviewed group the available space per person and the available crew per person on board is relevant for the luxury experience.

To make it possible to compare the existing cruises with each other, the amount of crew was divided by the amount of passengers on board (Crew per passenger = c/p). This for both private charter and cruises.

For the cruise vessels and yacht charters the available square meter per person was determined, using the available information of the cruise agencies and design drawings. By dividing the total available square meters with the total amount of passengers on board comparable parameters were created (Square meters per passenger = m^2/p).

The areas are divided in outside and inside public spaces. The inside public spaces were divided in entertainment, bar / lounge, sport and restaurant.
The private spaces were divided in 2 person cabin inside and outside, single cabin inside and outside and penthouses.
For these categories the m^2/p was calculated as well.

2.1.(c) Combining luxury and cost price

As expected, after analysing the data, the m^2/p and the €/p.p.d. are both higher on the charter yachts. Probably our concept will be between the values of the cruise vessels and the charters. But as earlier stated the cost price of the charters are too high and the m^2/p of the cruise vessels are too low. To determine the desired comfort of the cruise vessel the generated data will be combined and further analysed in the next chapter.

2.2 DETERMINATION OF THE NEEDED COMFORT

By combining the luxury and the cost price the desired comfort was described. The luxury was descripted as m^2/p and the cost price as €/p.p.d. The comfort parameter is created by combining these parameters into €/ m^2 p.p.d. By comparing the comfort parameter with the m^2/p of each accommodation category, results in the needed space. For the most spaces graphs were created. In this paper ,however only 2 are displayed and discussed.
Looking to figure 1, we see that there is a gap (blue ellipse) between the cruise vessels and charter yachts. The desired parameters for the new concept will probably be in the blue area of the figure below.

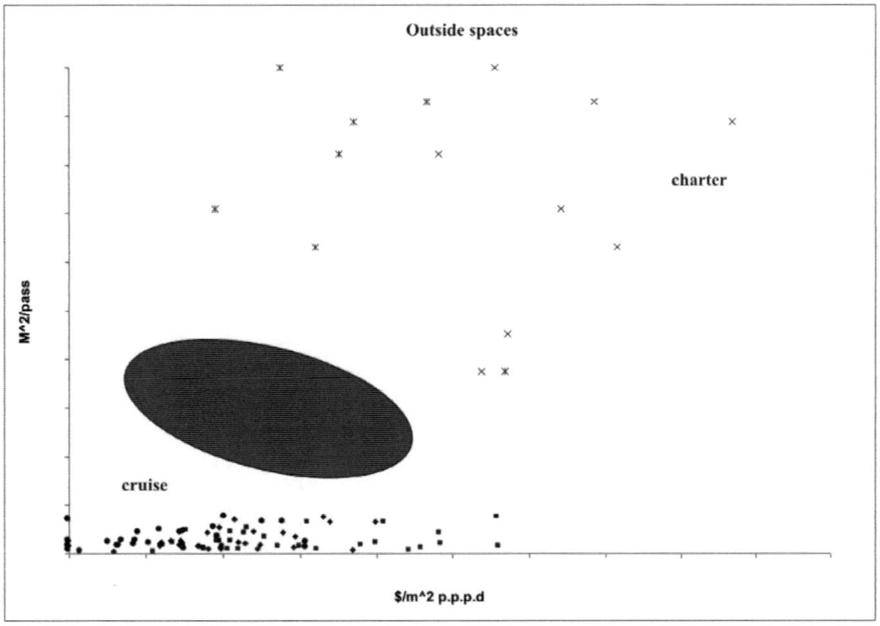

Figure 1: Defining the needed outside space.

Looking to the second figure we see that there is also a gap (blue ellipse) between the cruise and charter vessels in respect to the cabins. Here as well we may assume that in this area the desired parameters for the cruise vessel can be found.

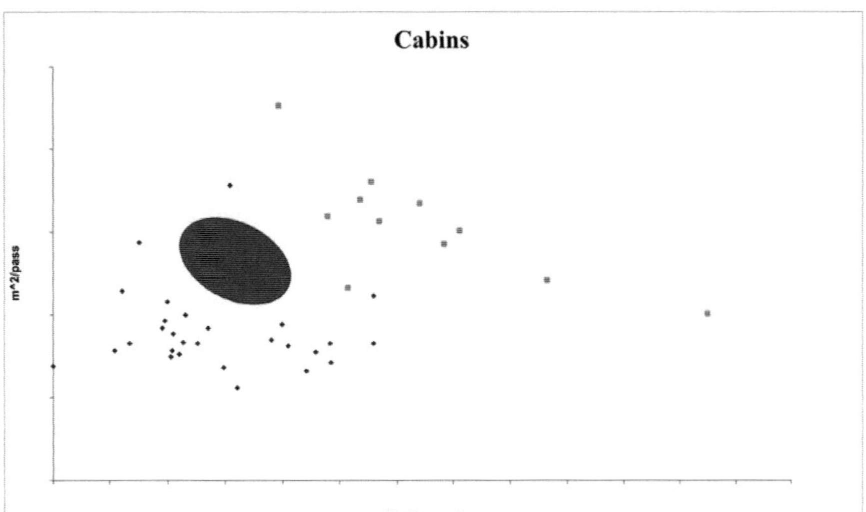

Figure 2: Defining the needed private space.

While studying the generated data, based on comfort, listened in different figures and graphs, the design criteria of the concept cruise vessel is defined.

3. DEFING THE DESIGN CRITERIA

The two aspects having impact on the desired design criteria of the cruise vessel are comfort and the sustainability in relation to the accommodation. In chapter 3.1 the design criteria and the needed parameters of the cruise vessel are defined. In chapter 3.2 the criteria of sustainability is described.

3.1 DESIGN CRITERIA ACCORDING THE ACCOMODATION

The square meter per passengers for the different accommodation type is determined according the graphs and figures, as listened and discussed, in the previous chapter.
The passengers should be between 40 and 110. The studied data (not shown in this paper) told us that the yacht charters had a maximum of 40 passengers on board and the cruise vessel a minimum of 110 passengers on board. The user group will be between the charters and the mass cruises. By multiplying the amount of passengers and the needed square meters per passengers the total needed area of each accommodation is determined.

To determine passenger and length ratio of this concept was determined with the so called trial and error method. We based this method on the idea to approach the data form different angles i.e. $L_{.o.a}$ (Length over all), amount of passengers and V_{cruise}.

The V_{cruise} angle results in the $L_{.o.a}$, the amount of passengers angle results in $L_{.o.a}$ and the $L_{.o.a}$ angle results in the amount of passengers. Figure 3 shows the results of the iteration process used for determining the needed parameters.

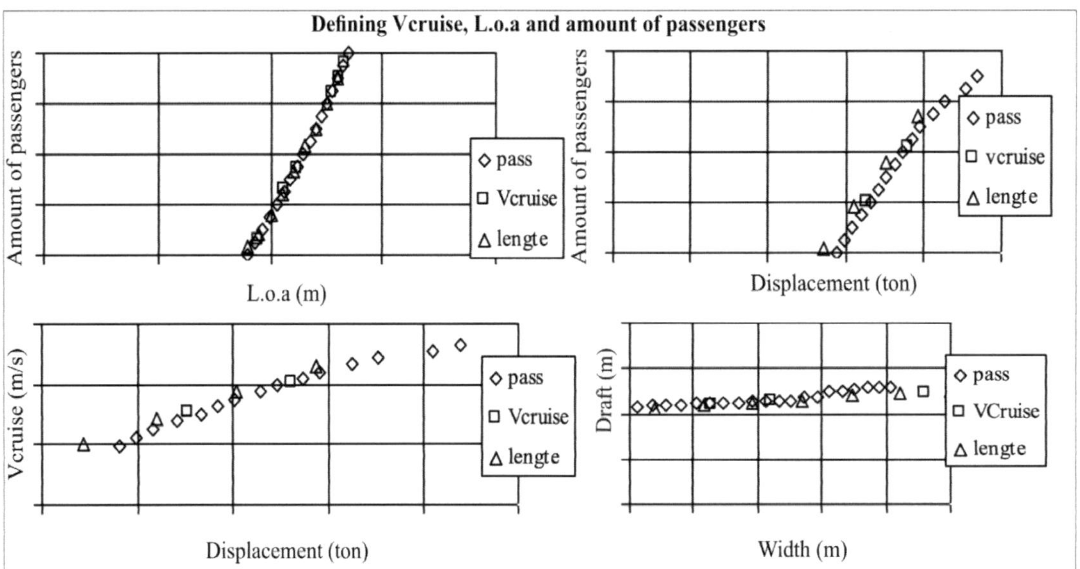

Figure 3: Defining the parameters of the cruise vessel.

The amount of passengers, V_{cruise}, Displacement Width and draft are determined with the figure (figure 3) above.

3.2 DESIGN CRITERIA ACCORDING TO THE ENVIRONMENT

Due to the awareness of the need to reduce the consumption of natural resources and pollution of the one planet we all share, sustainability is an important aspect in the design process. Everyone is convinced of the necessity to reduce the use and spoilage of the natural sources available.

A linear model from resource extraction, production, distribution, consumption and disposal causes many undesirable side effects.

A solution to reduce today's consumption or usages of the sources can be found in models that are based on the metabolism of ecosystems. These complex systems are comprised of dynamic feedback loops and are modelled according to the circular nature of natural processes. The paradigm shift influences architectural design and the building industry with its major impact on the environment due to the large amounts of materials that are used. The materials 'flow' through the building: meaning each part, each material has a life-cycle that starts with the extraction of the raw materials, continues through a chain of processing steps, is shipped several times and installed in the building. After remaining for a certain amount of time the materials are removed and either discarded or recycled. Each step along the line of building, using, and discarding/recycling will cause some kind of environmental impact.

Some materials and components complete the above listened cycle faster than others. Replacing a certain defective part can trigger a cascade of replacing other parts. This phenomenon is described by Duffy and formalised by Brand [2,3] as shearing layers where those layers with different speeds cause friction, i.e. waste and therefore pollution. This leads to the design concept of pace-layering, where components with different life-cycles do not hinder each other, allowing for a high level of adaptability. Extending the concept of different pace in the built environment beyond materials by including consumables and energy enables us to capture the building as a system that can be evaluated in terms of environmental impact throughout its whole lifecycle (European building regulations do not yet evaluate building materials and energy consumption in a unified set of rules).

By isolating each component with different lifespan, each component can be replaced or repaired separately without having an impact on the others items around them [2,3].

In chapter 3.2 (a) different aspects of quantifying the total impact will be discussed. The strategy for reducing the environmental impact is discussed in chapter 3.2 (b)

Environmental impact becomes a quantifiable measurement for the (a certain) quality of the building and can be used for comparing different design options. This is especially interesting with data-driven parametric models that can be programmed for self-evaluating and adapting.

3.2 (a) Quantifying the environmental impact

Environmental impact, also referred to as environmental footprint can be calculated and expressed within different reference systems by means of indicators quantifying the footprint. Over time a variety of indicators were developed, each with their specific emphasis on certain environmentally consequences or evaluation methods.

One indicator is called Material Input per Service Unit (MIPS) [4] and is focusing on the material input (the backpack) of a product or service unit. It is about the amount of material needed for production. The output, the so called after life of the product, is stated equal as the input.

An more specific and detailed/accurate procedure is the so called ReCiPe [5]. ReCiPe takes a multi-layered approach and expresses impact based on 3 scenarios of technological progresses.

The 18 different implemented factors have all impact of the possible product and environmental damage. Due to the complexity and time consuming describing of each factor, this procedure is not involved in the design process of this cruise vessel.

A widely adapted approach is the concept of carbon dioxide equivalent, CO_2e, comparing the negative impact on the environment with the amount of CO_2 with a comparable negative impact. Another common method is environmental cost accounting with a value expressed in a currency unit to indicate the amount of money that would be necessary to counteract the negative impact.

In a lifecycle analysis (LCA) the impact is calculated based on precompiled lifecycle inventories that are either provided for free by publicly or privately funded research (IDEMAT), are maintained under governmental auspices (NMD) or are available as commercial product (Eco invent). In order to receive comparable results that inform the design process, the desired performance is specified.

The four relevant factor of the LCA are:
- Mass and specific environmental impact of all materials (extraction to raw material)
- Processing steps

- Transport distances
- End-of-Life scenario

3.2 (b) Impact driven design choices

Based on the four factors as listened in previous chapter, a series of strategies can be employed to minimize the environmental impact during the design phase.

One strategy is to reduce the amount of material. when the component is crucial for its function or for the overall quality (and cannot be avoided), different approaches can be considered to alter the design.

The different approaches can be:

Reducing the dimensions of the applied materials. This will be an option as long as it does not influence the impact of the quality and lifespan.

Mitigating waste by considering the standard dimensions of the applied material is another option. By adjusting the size of the design grid on the standard dimensions of the used material, the rest material (waste) will be reduced.

Reducing dead weight by stripping all unnecessary material will be applied when allowed. For example the weight (mass) of an I beam with web and flanges can be reduced by perforating some parts of the beam.

Using material properties in the most efficient way by changing the topology of the structure is an efficient method used in architecture. It is based on the principle by employing thin structural members that are tensioned rather than being compressed. It can reduce the mass because buckling is no longer jeopardize the structural integrity of the part. This principle is applied by membrane structures for large roof surfaces of buildings.

Changing the geometry of the component and the design for accommodating the strength/stiffness may have impact on the component. Instead of adding additional parts to stabilize a flat surface (and using more material), the shape of the surface itself can be altered. A widely used example is the applied in the die forming process of sheet metal for car bodies.

Many of the approaches mentioned are greatly benefiting from computer-based design tools that are available and associated manufacturing techniques.

Reducing the specific impact by considerate material choices a decision tree cab be utilized to ensure a minimal environmental impact. Reclaimed material is the first choice to be reused when possible. In some cases it has to be modified before being applied. As second option bio based materials are taken into consideration The third option are new, recycled materials that rank lower because of the high degree of manufacturing necessary. New materials are applied only when no other viable option could be found in time.

Reduce energy related impact by minimize transport and additional processing steps. Using locally available materials reduces the need for long distance transport and has the additional benefit of employing local skills and thus creating a local style comparable to the traditional way of building. That is especially true for architecture and artisanal craftmanship but could inspire industrial production as well.

As processing can be energy intensive it should be reduced to and should prefer less energy demanding methods. In addition, processing can release undesirable or toxic substances that can be harmful for the workers or the environment in general. International standards for handling hazardous materials should be applied even if local codes are less strict.

Reducing impact by prolonging the useful lifespan of components or materials is an efficient way to reduce environmental impact since it mainly depends on design decisions. On the level of the building component (i.e. a section of a drywall system) it is desirable to create an ecosystem of interchangeable building components that can be removed and reinstalled non-destructively. The modularity of building components is a key to ensure that no materials are wasted when the functional program of building is updated. Detachable connections also play a key role in the End-Of-Life scenarios for materials. Only if materials can be separated without degradation of the quality, they are fit for reuse or recycling.

3.2 (c) Materials

Based on the reference frame of circular design principles introduced above the materials applied in the hospitality part of the design will be explained. Those examples have proven their feasibility in circular architectural designs over the last decade. Due to the technical complexity of the matter, the structural components of the ship are not within the scope of this paper, although the design considerations could be equally applied.

Another relevant aspect of material choice are fire safety considerations. The materials must meet the legal requirements of European building codes for ignitability, smoke production and emission of flaming particles or droplets. In addition the fire resistance and durability have to be sufficiently checked for the use case.
Risks can often be mitigated by appropriate material choices, additional treatment or the application of a covering layer.

4. SETUP THE DESIGN

The determined accommodation area in the previous chapter 3.1, is used to create an accommodation diagram (see figure 4). However the square meters per person per day is the base of the design of the cruise vessel the interior and other aspects are as relevant as well.

The design concept for the new cruise vessels is focused towards the interior design of the ship. Taking into account the user's wishes and demands, the activities on board and a block diagram of the accommodation form the basis of the vessel design [6,7,8,9].

According the extensive research of the vessels, observations and interviews the layout of the cruise vessel is established. Figure 4 shows the block diagram of the cruise vessel each block can be further divided into specific detailed area. This process continuous until the entire vessel is finalised.

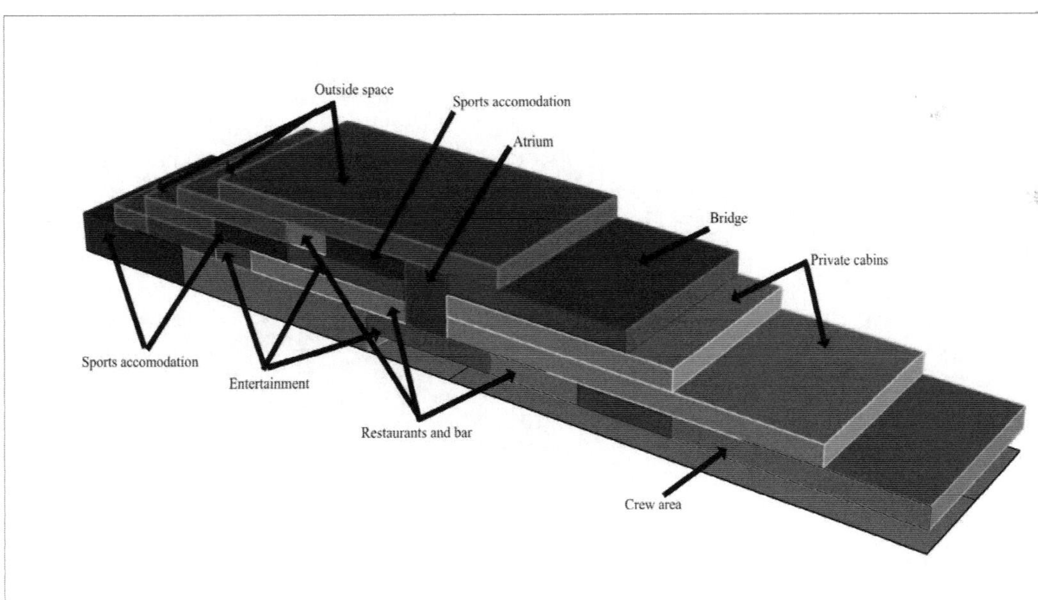

Figure 4: Block diagram of the vessel.

The character and the main use of the area plays an import role for the location. A cinema (green coloured) for example doesn't need to have windows and can be placed below the waterline. A bar or restaurant however should have windows for enjoying the view. The reception should be near the entrance (purple coloured). The cabins should all have a window to give the passenger a nice view and the desired perception of luxury. The location of the cabin (cyan coloured) should be as far as possible away from the working area and the restaurant / bar area (yellow coloured). This should reduce the sound and vibration levels in the cabins, which is needed to create luxury.

The outside space (red coloured) is on the top deck covered with solar panels, mainly to create solar power (discussed later in this paper) and to create shadow.

The sports /recreation area (blue coloured) is situated partly below the waterline and partly above. On the first deck the sports area is situated partly inside and outside. Passengers can sport depending on the weather in or outside.
Although the accommodation is created as spacious as possible, unfortunately the available space is limited. Luxury should be created by using other features as well. One of the features is sustainability as discussed in chapter 3.

4.1 EXTERIOR DESIGN

After finalizing the block diagram and the design criteria with the needed parameters the exterior of the cruise vessels can be created. One main aspect of comfort is the climate control of the different areas. In the winter the sun will help to heat up the space and is needed to shine into the accommodation, in the summer however shadow is more required. Besides the sun also the need of fresh air can differ per season. These two aspects are taken into account by designing the exterior of the vessel.

The top deck is covered with solar panels to generated electricity. The location of the windows and the location of the needed ventilation grids of the accommodation influences the exterior of the vessel. Both of these have an impact of the power consumption of the vessel. As mentioned earlier the propulsion energy is not taken into account in this paper and therefore only the accommodation will be discussed.
The plants situated on the deck, are the so called reed beds and will be discussed in chapter 4.3. The air intake and light tube are contributing to the applied systems as discussed in chapter 4.3.

Figure 5: Exterior of the cruise vessel.

4.2 INTERIOR DESIGN

For the interior design alternative materials and concepts, to reduce the impact of materials (in the hospitality part), are applied in the design. Those considerations are derived from circular design practices that became relevant in architecture over the last decade. However due to the technical complexity of the matter, the structural components of the ship are not within the scope of this paper.

In addition to select materials with a higher degree of circularity, the interior is designed based on a system of interchangeable components. By installing the elements the detachable connection is important to easily dismount /disconnect the material. Glued or welding connections should be avoided. In this case it is able to swap elements when needed.

Implementing second hand materials and material created from recycling and waste, the footprint is reduced. The environmental impact is reduced by using local materials when present. Reducing the transport of materials has a big impact of the environment as discussed in chapter 3.2. (b).

Figure 6: Interior of a cabin with a view.

4.2 (a) Walls of the cabins

The non-loadbearing parts of the walls of the cabins, as shown in figure 6, consist of 3 parts the beams, isolation and cladding. These three components can be easily separated during dismantling at the end of the life cycle of the cruise vessel. The top layer of the wall the' so called' cladding of the cabins are the cut out steel plates. These plates are the leftovers of the production process of products. After cutting out the shape of the plates needed the rest of the plate will be like this. Instead of throwing away these plates we use them as decoration of the inner walls of the cabin (figure 7 shows the materials used for the wall).

For the insulation sheep wool is used. Sheep wool has a good insulation value and has a good fire resistance. Especially in the Netherlands this wool is after shaving the sheep thrown away. By applying wool for insulation the garbage will be reduced and the cost price of the vessel will be reduced since the material is quite cheap.
The curtains are as well fabricated from wool. Although it is tried to reduce the amount of elements (the less elements the lesser the impact on the environment) in the cabin, the use of curtains will be still essential due to the effect on the acoustics. Besides sustainability comfort still plays a big roll and acoustics will contribute to the comfort of the cabin.

For the beams second-hand wood will be applied (see the left topside of figure 8), this wood is taken out of buildings during the demolishing phase and are tested and certificated before applying for the second time.
In the so called wetted spaces like the shower, toilet and bathrooms other cladding material is used. The material applied in these areas are the so called reused HPL (High Pressured Laminate). These plates can be decorated by milling a pattern in the upper layer, besides decoration it will have an anti-slip function as well and can be used on the floors.

Figure 7: The composition of the wall.

4.2 (b) Floor and ceiling

Floating screed is a common method to uncouple the floor surface from the structural slab. Three layers are distinguished by function; insulation against heat or transfer structure-borne noise and vibrations, a loadbearing and load-distributing layer and an additional technical systems (i.e. pipes for heating and cooling etc) layer.

The insulation layer, mostly not visible, consists of reused cover or top ceiling panels. During the usages of the these panels the cosmetics will decrease while the material quality stays more or less the same. Since the insulation is not visible cosmetics are not relevant in this case.

The load distributing layer consists of reused HPL or plywood, in cases when needed this layer can be double to comply to the regulations.

The top layer consist of recycled material. Due to the visibility of this layer cosmetics plays an important role. Recycling is in this case a better solution than reusing material and components. In the so called wetted spaces second-hand HPL plates will be used as top layer. These plates are sanded to create the desired anti slip. On the right side of figure 8 HPL is shown with different decoration aspects.

For the ceiling the same procedure is used. Besides insulation for structure borne noise, insulation for heating/cooling and lighting is added. The applied constructing consist of a suspension system, a structural grid, insulation material, the cladding and finishing layer.

Figure 8: The applied materials.

4.2 (d) Furniture

The furniture is made by local companies and consist of recycled materials. If it is not possible to use recycled material wool other biological material is applied. By reducing transport and the usage of new material the footprint is reduced. By constructing and designing the furniture using the principles discussed in chapter 3.2 impact can be sufficient due to the amount of furniture distributed in one cruise vessel.

4.3 ENVIRONMETAL DESIGN (TECHNICAL SYSTEMS)

The energy to power the technical installation is also causing an environmental impact. This paper does not cover the propulsion system where due to the massive power consumption the highest benefits can be achieved. Research is conducted in the field (optimizing drive train, fuel, reducing drag) and is already taken care off by other companies.

Providing services for the hospitality on board the vessel consumes roughly 12% of the average power produced on board. By exploring alternative systems like they are already implemented in buildings the consumption can be reduced. The energy input for the water system can also be reduced by bringing back the amount of fresh water that is produced.

One way to make systems more sustainable is to substitute carbon based energy by renewable energy that is generated regeneratively by the sun or the wind. Like shown in figure 5 the installed solar panels on the rooftop deck.

Although the energy provided by the sun can cover a large share of the electric consumption, it remains difficult to supply that energy permanently. In order to fulfil the power demand batteries should be applied, however due to the weight and the needed space this option should be avoided when possible.

Because the demand and the supply of power (in this case electricity) is not always synchronized, it would be necessary to store large amounts of energy. As battery technology improves constantly it might become more likely to easily store sufficient power. Besides the discrepancy in the supply and usage of the electricity there is also a discrepancy with the water supply. The volume of the buffer limits the availability for providing fresh water continuously on demand. Guarantying the sufficient fresh water or electricity supply, systems with large over capacities will be needed to be installed. Having the sustainability in mind over capacity is not the desired solution.

4.3 (a) Water system

Currently fresh water is either bunkered at shore or produced on board with water makers. In order to save on space for tanks and energy for the purification process, the water system could be enhanced with rainwater catchment and alternative onboard wastewater treatment.

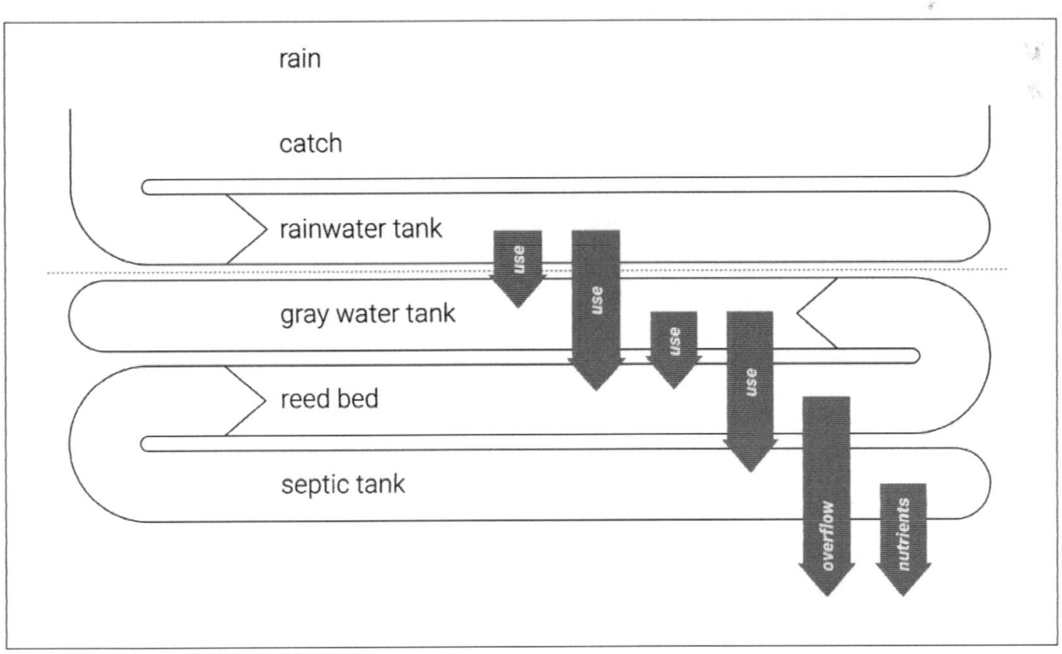

Figure 9: Water reuse system.

Rainwater can be collected on deck and in the reed beds (unfortunately only the use of the reed beds isn't sufficient). As currently used in systems in buildings, first the water is used to flush the collecting surfaces and then discarded. Water is successively channelled to tanks, passing through filters to prevent debris from entering the system.

Depending on the intended use, the collected water can be purified by carbon filters and stored. The dimensions of the tank ensures that water does not remain in the tank for an extended period.

Purified rainwater can be used in the hospitality section of the vessel for cleaning purposes and for flushing the toilets. Although it is possible to provide potable water from rainwater, this is currently not applied (due to the regulations).

Once rainwater is used, it enters the black- and greywater treatment system through the septic tank or the reedbeds (shown in figure 5).

Black water is transferred to a septic tank where solid matter will settle on the bottom. The sludge from the septic tank could supply a biogas unit onboard for generating electricity. The fluid fraction from the septic tank is transferred to reedbeds on deck where it remains underground (in the soil of the reedbeds) and bacteria on the roots of the plants takes care of the nutrients in the water. In addition to the water cleaning function, evaporation from the reeds and other plant species will help to cool the air in warmer climates.

As a final step the water can be aerated in open cascades. That raises the oxygen level which will help to preserve the greywater.

It is ensured that (rain)water that was used once, will not enter the hospitality section of the vessel again. Greywater is used for technical purposes only.

4.3 (b) Ventilation and air-conditioning

Natural ventilation will not substitute the vessel's air conditioning infrastructure, but will help to reduce its size, will simplify it and make it more energy and cost-efficient.

As long as the outside temperature and humidity allows for direct ventilation, fresh air enters the cabins directly from outside. For the corridors air is supplied on the shortest route from outside.

The used air is exhausted from the bathrooms and toilets of the cabins. Air in the corridor is extracted via the atrium a alongside the light tube) or ventilation shafts (air intake figure 5). Shaped hoods on the upper deck employ the Venturi effect to create a pressure differential to support the desired airflow.

If outside conditions demand heating or cooling, air is treated in modular units that supply a cluster of four cabins. In this case the units needed can be compact and the climate can be individual controlled per cabin. The energy required is supplied form a centralized installation by liquid media.

Additional warmth in the bathrooms and spa areas can be provided by integrated infrared panels. Infrared panels guarantee high level of comfort while being energy efficient.

5. CONCLUSIONS

Developing a new holistic cruise vessel concept fulfilling the demands of the so called early cruisers and the sustainable needs is an creative task combining aesthetical technical and sustainable components. Lowering the environmental impact and the ecological footprint doesn't interfere with the expected luxury on board.

Using the observational and processed data a new definition of comfort and luxury is created. Processing different parameters in a dynamical process results in the design criteria of the cruise vessel. These criteria fulfil the needs of the specific group of early cruisers.

Different holistic design approaches for developing and implementing sustainability in the design process of this cruise vessel are used. Implementing the different design aspects and methods during the design process is possible without interferences and clashes.

The new cruise vessel concept, as developed, in this paper contribute to a better image of the cruise sector and to a new market potential expending in the near coming future.

6. ADDITIONAL INFORMATION

In this paper a new holistic concept development of a luxurious sustainable cruise vessel is described. However most of the applied aspects are used in the architectural environment, before applying it in the shipbuilding industry, attention should be paid to the classification rules and demands. Because of some interest from the industry not all data and parameters are shown and shared. Please feel free to contact the author for more information. IJgensinnig will be pleased to help developing this cruise vessel with other stakeholders and to help them with its experiences and knowledge.

7. REFERENCES

1. BUIJS J., VALKENBURG R., 'integrale product ontwikkeling', *TU Delft*, 2000

2. BRAND, S., 'How Buildings Learn', 1994

3. BRAND, S., 'The Clock of the Long Now', 1999

4. RITTHOFF., et al, 'Calculating MIPS', *Wuppertal Institute,* 2002

5. HUIJBREGTS., et al, ReCiPe 2016 v1., *RIVM,* 2017

6. MONCHY, M M de., 'From landlubber to seaman', *Marine Design 3-4- September 2014* Coventry, UK

7. MONCHY, M M de., 'Creating a new distinctive identity for a range of offshore vessels using an industrial design approach' *Marine Design 3-4- September 2014* Coventry, UK

8. NELSON, T., 'The potential for surface disruption techniques in the aesthetic evolution of superyachts', *Marine Design 3-4- September 2014* Coventry, UK

9. JABTANOM, A., PHORMTAN, T., NAZARAOV, A., 'Design development and construction of 65' catamaran' *Marine Design 3-4- September 2014 Coventry*, UK

10. BASTIEN, C., MALI, L., 'A light weighting methodology of a trimaran high speed cruise logistic ferry' *Marine Design 2-3 September 2015* London, UK

8. AUTHORS BIOGRAPHY

M M de Monchy MSc holds the current position of concept designer at design studio IJgensinnig, situated in Rotterdam.

Due to his background as a naval architect BSc and industrial Designer MSc and Business science BSc, he creates new holistic concepts fitting the new market possibilities. The Author is responsible for developing new concepts driven by market research. Besides developing new concepts, developing the exterior and interior of vessels and houses according clients wishes is also his task. Designing from the user's perspective is the speciality and main core business of M.M. de Monchy. 'The design should fit like a personal coat' is one of his most quoted expression. The Author's previous experience includes holistic concept designer, stylist within different shipbuilding companies.

Frank Feder graduated from Bauhaus University with a diploma in architecture. In 2008 he founded FF-arch in Rotterdam, focusing on circularity in architecture, regularly contributing to pioneering and award-wining projects of Superuse Studios. In this paper, Frank introduces base concepts of circular design in architecture and instruments used to evaluate the environmental impact and explores how these concepts can be applied in the design of the hospitality aspect of the vessel design.

A COMPARATIVE REVIEW OF ALTERNATIVE FUELS FOR THE SEARCH AND RESCUE CRAFT

N Ghahramani, **A J Murphy**, **S D Benson** and **R W Birmingham**, Newcastle University, UK
N Chaplin, **S Johns** and **V Limbrick**, Royal National Lifeboat Institution, UK

SUMMARY

At the 72nd Marine Environment Protection Committee (MEPC) meeting in 2018, the International Maritime Organization (IMO) adopted a strategy to reduce the carbon footprint of global shipping by 50% by 2050 compared to the 2008 baseline. The agreement exemplifies the urgent global need to decarbonise shipping and, by extension, all forms of marine vessels. Consistent with this, the RNLI, with a significant role in the deployment of maritime search and rescue services across the UK and the Republic of Ireland by providing more than 400 lifeboats, has set a long-term ambition of becoming a zero-carbon organisation for Scopes 1 and 2 emissions by 2050. As part of their efforts to achieve this, they are working with Newcastle University on research which aims to identify alternative low and zero-carbon powering systems for the RNLI's vessels. In response to the anticipated demand for engine systems to run on low-carbon fuels, engine manufacturers and some maritime organisations have started developing alternative energy systems. However, at this stage, they are typically aimed at large vessels; therefore, limited activities for decarbonising small craft have been provided. There are often weight and volume constraints as the main challenging factors in designing new systems for small craft. Newcastle's research has reviewed various energy systems towards low and zero carbon emission for RNLI craft. Hybrid systems including fuel cells and batteries will be the focus of this paper. A volume and weight comparison, carbon emission and cost comparison of alternative fuels, and safety considerations are the emphases of this study. A comparison framework to evaluate alternative fuels for powering fuel cells specifically aimed at small boats is proposed.

NOMENCLATURE

ALB	All-weather Lifeboats
ICE	internal combustion engines
ILB	smaller inshore lifeboats
FC	Fuel Cells
LCV	low calorific values
MDO	Marine Diesel Oil
SAR	Search and Rescue
SFC	Specific Fuel Consumption

1. INTRODUCTION

The International Maritime Organisation (IMO) adopted an initial strategy in 2018 for large merchant shipping aimed at reducing Greenhouse Gas (GHG) emissions per transport work by at least 40% by 2030 and 70% by 2050 compared to 2008. The short-term plans established in 2018 focused on improving energy efficiency and conducting research on alternative fuels and technologies for reducing GHG emissions until 2023. The mid-term plans (2023-2030) involve the implementation of alternative fuel programs and emission reduction mechanisms. It is encouraged to share these initiatives with other corporations for learning purposes. And finally, the long-term plans (2030-2050) aim to facilitate the adoption of emission-reduction technologies and pursue the development and provision of alternative fuels [1].

The long-term measures are to peruse the development and provision of zero-carbon fuels for decarbonization in the second half of the century. The Marine Environment Protection Committee (MEPC) is set to adopt the updated and revised strategy during its 80th session scheduled to take place from July 3 to July 7, 2023. [2]. However, the current focus of marine decarbonization strategies is primarily on larger vessels, and as a result, the role of small craft is often overlooked.

Small craft, such as leisure boats, fishing vessels, and lifeboats, play a significant role in the maritime industry and contribute to overall emissions. By considering the unique characteristics and requirements of small craft, tailored decarbonization plans can be developed to ensure that all segments of the maritime industry contribute to a sustainable and low-carbon future.

The Royal National Lifeboat Institution (RNLI), as a Search And Rescue (SAR) operator in the UK and Ireland, is committed to mitigating the environmental impacts of carbon emissions and landfill waste, with the goal of achieving a zero-carbon footprint [3]. The RNLI operates a fleet of over 400 lifeboats across the UK and Ireland, ranging from All-weather Lifeboats (ALBs) to smaller inshore lifeboats (ILBs). The selection of a particular type of lifeboat is determined by factors such as the geographical location and the nature of the rescue mission.

The current fleet of the RNLI is powered by diesel and petrol internal combustion engines (ICEs). Fossil fuel consumption within the RNLI encompasses both road transport and vessels. In 2021, the RNLI consumed a total of 2,048,017 litres of Marine Gas Oil (MGO), with 872,865 litres allocated for road fuel, including some purchases for ILBs. In line with the global trend towards achieving zero carbon emissions by 2050, it is imperative to replace or adapt the current propulsion systems with alternative fuels, as outlined in the short and long-term plans designated until 2050.

Several factors must be considered for the implementation of alternative propulsion systems, including safety precautions and associated risks, Capital and Operational Expenditure (CapEx and OpEx), the amount of GHG reduction considering the fuel production process, and the development of required infrastructure and training programmes to ensure familiarity of the staff with the operation and maintenance of the new systems.

The operational characteristics of the RNLI's boats pose a particular challenge for this endeavour due to their compact size and the complexity of their mission profiles, which often involve high-speed trips in a variety of different weather conditions.

Reducing GHG emissions can be achieved through two means: A) using low or zero-carbon fuels, and B) adopting new technologies to improve the technical efficiency of machinery. Yet, the use of alternative fuels has shown to be significantly more effective in carbon reduction. With respect to this aim, this study compares the application of Fuel Cells (FCs) powered by low and zero-carbon fuels, as a viable alternative to ICEs powered with Marine Diesel Oil (MDO).

2. LITERATURE REVIEW

SAR craft missions necessitate a rapid and powerful response to achieve the desired situation response times. RNLI lifeboats are equipped with the required power to achieve vessel speeds up to 40 knots (25, 35, and 40 knots for ALBs, B-class, and E-class respectively). The RNLI operates a diverse fleet of lifesaving craft, including All-Weather Lifeboats (ALBs) with propulsion power ranging from 1300 to 3200 hp, as well as Inshore Lifeboats (ILBs) with propulsion power varying from 50 to 900 hp, depending on the specific boat type. The RNLI lifeboat fleet has been developed to provide a range of vessels with capabilities matched to the required operational profiles. Tactical ILBs, capable of speeds ranging from 25 to 35 knots are deployed to cover close inshore areas where the risk of casualties is higher. ALBs are located and deployed to handle larger vessel incidents and casualties. This requires extended range, weather resilience, and enhanced capabilities to cover areas up to 100 NM offshore with boat speeds up to 25 knots.

In this context, the focus of this study is on exploring the use of batteries and fuel cells as low and neutral-carbon power solutions for high-speed boats. The context of this paper is to explore the development of fuel cell technology as a potential future power solution for lifeboats, whilst also acknowledging that other alternative fuel solutions may ultimately be adopted by the RNLI.

2.1 FUEL CELLS (FC)

Fuel cells convert reactant chemical energy to electrical energy, resulting in low to zero carbon emissions depending on the fuel used – as a wide range of fuels including fossil fuels as well as carbon-neutral fuels can be used by different types of fuel cells. In the marine industry, there is a growing trend towards the adoption of fuel cells due to their high efficiency and energy density, enabling vessels to have longer endurance, as well as the ability to generate power with zero emissions. Table 1 presents some characteristics of the most used fuel cells in the maritime industry.

Fuel cells are classified based on their electrolyte types, fuel type, and working temperature [4]. In the maritime industry, four types of fuel cells are commonly used due to their energy efficiency, power capacity, and sensitivity to fuel impurities [5]. These fuel cell types are discussed in the following section along with their operational characteristics, and some examples of their application in marine projects.

Hydrogen, ammonia, natural gas, and methanol are identified as potential low and zero-carbon maritime fuels for fuel cells. Fuel cells are primarily considered for small-sized boats or passenger ferries due to their limited power output [5]. However, larger ships view FCs as Auxiliary Power Units (APUs). Some studies suggest that as FCs continue to mature industrially, they may achieve higher volumetric power density compared to diesel engines [6].

The United States Coast Guard (USCG) conducted a feasibility study on replacing the diesel generators of the Vendicator diesel-electric vessel with FCs. This study involved the installation of four modules of 625 kW MCFCs, which were fuelled by low-sulphur oxide distilled marine fuel (F-76). The vessel's stability and performance remained unaffected. However, to accommodate the larger dimensions of the FCs compared to the diesel generators, non-structural side plates in the generator rooms had to be removed [7].

Low-Temperature Polymer Electrolyte Membrane Fuel Cell (LT-PEMFC) can be fed with pure hydrogen. However, pure hydrogen has a low volumetric density, which makes it unsuitable for long-distance shipping applications. LT-PEMFCs are the most developed type of fuel cells and operate by consuming hydrogen and oxygen to produce electricity, water, and heat $[O_2 + 2H_2 = 2H_2O + heat + electricity]$. Most ongoing projects in the maritime industry utilize LT-PEMFCs with hydrogen as the preferred fuel. This is especially suitable for sailing services with relatively low energy requirements, as it allows for the addition of bulky fuel storage onboard. LT-PEMFCs typically operate at temperatures ranging from 65 to 85 °C and have limited tolerance to fuel impurities.

SF-BREEZE employed 120 kW PEM racks to demonstrate the feasibility of a high-speed passenger ferry powered exclusively by 1200 kg of liquid hydrogen (LH_2) stored in a Type-C tank located on the top deck. This fuel quantity was sufficient for two 50 nm round trips [8].

High-Temperature Polymer Electrolyte Membrane Fuel Cell (HT-PEMFC) is capable of tolerating reformate gas with a few per cent of carbon monoxide (CO) impurities. These fuel cells are also commonly fuelled by methanol (MeOH). HT-PEMFCs operate at higher temperatures, typically in the range of 120 to 180 °C, and they require additional start-up time compared to LT-PEMFCs. However, they have a higher tolerance to impurities compared to their low-temperature counterpart.

MS Mariella, a ferry with a passenger capacity of 2500, demonstrated the use of an HT-PEMFC with an electric power output of 60 kW, fed by methanol stored in a 6 m^3 double-walled tank [9].

Solid Oxide Fuel Cell (SOFC) is not sensitive to fuel impurities, making it suitable for use with high-density fuels rather than pure hydrogen. SOFCs operate at higher temperatures, typically ranging from 700 to 1000 °C. They are known for their high-power density and electric efficiency. SOFC can be directly fuelled with ammonia which opens up possibilities for utilizing ammonia as a fuel source in maritime applications.

METHAPU used methanol-powered 20 kW SOFC modules to supply electricity onboard a merchant navy vessel. The SOFCs in this project also incorporated heat recovery systems to produce fresh water and generate additional electricity. The surplus heat generated by the SOFCs was also utilized for conditioning purposes onboard the vessel [7].

Molten Carbonate Fuel Cell (MCFC) operates at temperatures between 600 and 700 °C and can be fuelled by various sources, including LNG, methanol, and other hydrocarbons, in addition to pure hydrogen. While working at high temperatures may have negative effects such as cracks and corrosion, MCFCs are known for their high efficiency and relatively low-cost catalysts.

The main drawback of MCFCs is their slow start-up time and limited flexibility in responding to power demand fluctuations. To ensure a steady operation, a combination of MCFCs with batteries is often necessary. MCFCs are commercially expensive compared to other fuel cell types. Moreover, their lower power density might be seen as a limitation for certain applications. Therefore, careful consideration is needed to determine if MCFCs are the most suitable choice based on specific requirements and constraints.

For low-temperature applications, LT-PEMFCs and HT-PEMFCs are recommended as fuel cell options, with hydrogen being the preferred fuel. LT-PEMFCs are suitable for applications where the energy requirements are relatively low and bulky fuel storage is feasible. HT-PEMFCs can tolerate more fuel impurities and operate at higher temperatures, offering advantages in certain scenarios. Hydrogen is a viable fuel for both types of PEMFCs.

For high-temperature applications, SOFCs and MCFCs are recommended, which can utilize various hydrogen carriers or renewable fuels as feeding fuel sources. A waste heat recovery (WHR) system can also be employed to enhance the overall system efficiency [5].

MCFC and SOFC systems generally have lower power densities, but higher energy efficiencies compared to PEMFC systems and therefore determining specific fuel consumption and storage volume is important to be considered when making decisions for the targeted boats.

Table 1. Key characteristics of fuel cells

FC type	**Fuel**	**Relative cost**	**Emission**	**Operation temperature (°C)**	**Fuel impurity sensitivity**	**Safety**	**Efficiency %**	
							Without heat recovery	**With heat recovery**
LT-PEMFC	H_2	Low	-	65 ~ 85	Medium	H_2 safety	50-60	-
HT-PEMFC	LNG, methanol, diesel, H_2	Moderate	Co₂ and NOx if carbon fuel used	120 ~ 180	Low	High temperature	50-60	80
SOFC	LNG, methanol, diesel, H_2, ammonia	High	Co₂ and NOx if carbon fuel used	700 ~ 1000	Low	High temperature	60	85
MCFC	LNG, methanol, diesel, H_2, hydrocarbons	High	Co₂ and NOx if carbon fuel used	600 ~ 700	Low	High temperature	50%	85

2.2 RECHARGEABLE BATTERIES

Lithium-ion batteries are regarded as the preferred type of rechargeable batteries used in electric vehicles due to their highest energy density (compared with other types of rechargeable batteries) with longer cycle life and the possibility to incorporate smart battery energy management systems for improved safety and performance [10]. Therefore, they are the only type of batteries considered for this study. The State of Charge (SoC) refers to the current level of energy stored in a battery relative to its maximum capacity.

2.3 FUEL CELLS AND BATTERIES IN HYBRID PROPULSION

The main difference between batteries and fuel cells lies in their operation: batteries are closed energy storage systems, while fuel cells are open systems that require a continuous supply of reactants. Batteries can be categorized as primary (non-rechargeable) or secondary (rechargeable). Some types of secondary batteries include lead-acid, nickel-cadmium, and lithium-ion, with lithium-ion batteries being favoured in the automotive industry due to their high energy density and specific energy.

Fuel cells have lower transient response capabilities compared to batteries, making their combination with batteries beneficial to address this limitation. Batteries can provide stability to fuel cells, reducing the strain caused by thermal cycling, especially in high-temperature fuel cells. Furthermore, during decreased load conditions, excess power generated by fuel cells can be used to charge the battery banks [5].

A Ragone plot, as shown in Figure 1(a), depicts the specific power versus specific energy of various storage technologies. Fuel cells have high specific energy but relatively low specific power, while batteries have higher specific power but lower specific energy [11]. This means that batteries are better suited for high-power demands but have limited endurance. Torqeedo offers fully integrated electric boat drive systems known as "Deep Blue Systems" which provide propulsion power of up to 200 kW and feature battery banks with a capacity of 1 MWh. As an example, the ZenPro 850 is a 5.8-meter Rigid Inflatable Boat (RIB) that is equipped with the Deep Blue 50R outboard and a 40-kWh battery [12].

In long-term operations, the continuous fuel supply to fuel cell systems gives them a higher energy density compared to batteries, as shown in Figure 1(b). Therefore, the report suggests that a hybrid fuel cell-battery system can provide both longer driving cycles and the required payload capacity. However, fully battery-electric boats may be suitable for certain lifeboat stations where shorter endurance is sufficient. This requires quantification with data specific to the RNLI fleet.

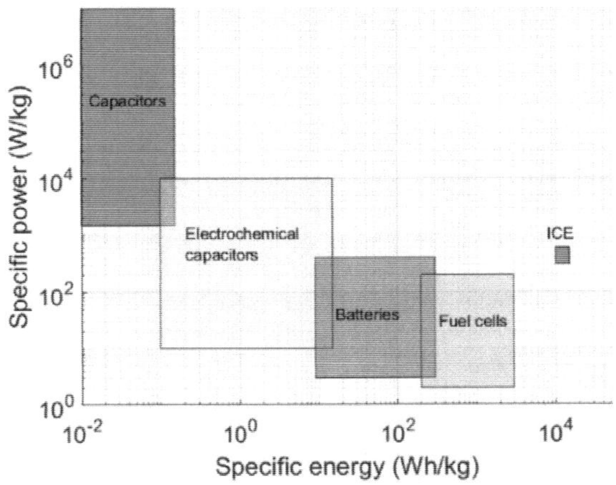

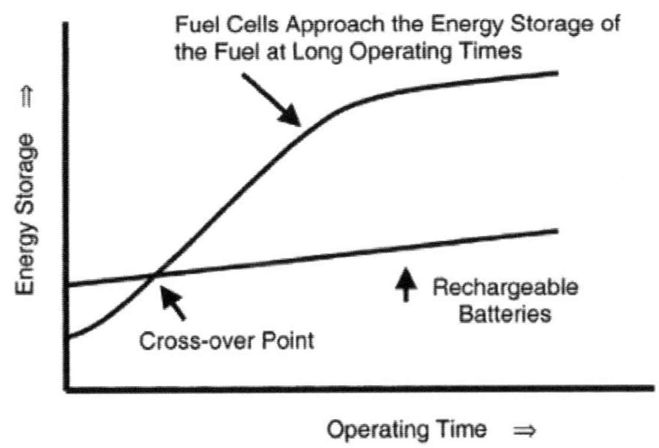

Figure 1: (a) Ragone plot [11] (b) batteries and FCs energy density comparison [13]

In conclusion, the following scenarios are suggested as alternative power solutions to the conventional systems:

- Utilizing fuel cells as a "range extender" to maintain the battery charge within the desired SoC limit, enabling longer operation without the need for frequent recharging.
- The combination of fuel cells and batteries in a hybrid power system allows for optimal power generation and management configuration, where the fuel cell serves as the primary power source, delivering the base power. During peak-power demands, the battery can be utilized to provide additional power, meeting the instantaneous high-power requirements. During periods of low power demand, the fuel cell can recharge the battery, ensuring its availability for future peak-power demands. This approach results in the reduction of large battery packs, which can be advantageous for small boats with limited space and weight constraints.

3. TECHNICAL CHARACTERISTICS OF FUELS

When considering the technical characteristics of alternative fuels, their infrastructure and onboard compatibilities must be considered. Infrastructural factors include the technological potential for fuel production and storage, transport and bunkering, while onboard compatibility involves volume, weight, machinery adaptability, safety, and reliability constraints. Section 2.1 compares some of the onboard capabilities of fuels including their storage energy density as well as the fuel cells' power density when being fed by the selected fuels.

3.1 ENERGY DENSITY

Energy density is a key factor that can impact vessel endurance limits. Alternative fuels with an energy density lower than MDO may require reducing the available space onboard for rescued survivors, more frequent refuelling intervals, or implementing a new hull design structure. Since the carriage of large numbers of survivors is a rare event, the most critical consideration would be the potential reduction in the endurance of lifeboats.

The gravimetric energy density of fuel (specific energy) is the available energy per unit mass of the fuel. This gives the energy content of fuel in terms of storage and handling. The volumetric energy density defines how much energy a system contains in comparison to its volume [14]. Table 2 presents some properties of fuels including their densities and low calorific values (LCV).

Due to the insulation required inside the tanks, the actual available space for fuel storage is reduced. Hydrogen has nearly three times the energy density of MDO on a mass basis, but on a volume basis, the situation is reversed. Compressed hydrogen requires large-volume composite vessels, which may have a greater impact on small boats with space constraints. To overcome these challenges, pressure vessels made of fibre-reinforced composites are employed, which can store hydrogen at up to 700 bar. Alternatively, cryo-compressed storage systems capable of storing hydrogen in liquid form with higher density have the potential to meet hydrogen storage targets.

Although compressed hydrogen cylinders are bulky, they are currently the practical choice for the transport industry. The United States Department of Energy (DOE) has set a target of achieving a volumetric energy density of 4.7 MJ/l in 2025 and ultimately 6.1 MJ/l, for compressed hydrogen used in light-duty fuel cell vehicles [15]. Liquid hydrogen is an alternative, but its high cost and boil-off feature raises questions about its suitability for widespread use.

Figure 2 and Figure 3 illustrate the volumetric and gravimetric energy density of alternative fuels relative to MDO, excluding and including the impact of storage systems on fuel density. This is particularly important when the fuel requires cryogenic or pressurized storage, such as hydrogen storage systems in various states. A relative volumetric energy density greater than 1 indicates that the fuel requires more storage volume than MDO, while a value less than 1 indicates that the fuel storage system is more compact than MDO. Similarly, a relative gravimetric energy density greater than 1 indicates a heavier fuel storage system relative to MDO, whereas a value less than 1 implies a lighter fuel storage system. It is notable that the ovals depicted in Figure 3 do not represent actual data but are derived based on nominal data. The purpose of these ovals is to provide a visual comparison with MDO storage systems.

Table 2. Physical properties of fuels

Fuel	CH2	LH2	NH3	MeOH	MDO	Biodiesel
LCV (MJ/kg)	120	120	18.6	20	41	38
Volumetric density (MJ/l)	4.6	8.7	12.7	15.8	35	32
Storage pressure (bar)	700	1	1 or 10	1	1	1
Liq. Storage temperature (°C)	20	-253	-34 or 20	20	20	20
density (kg/m³)	38	72.5	683	790	854	842

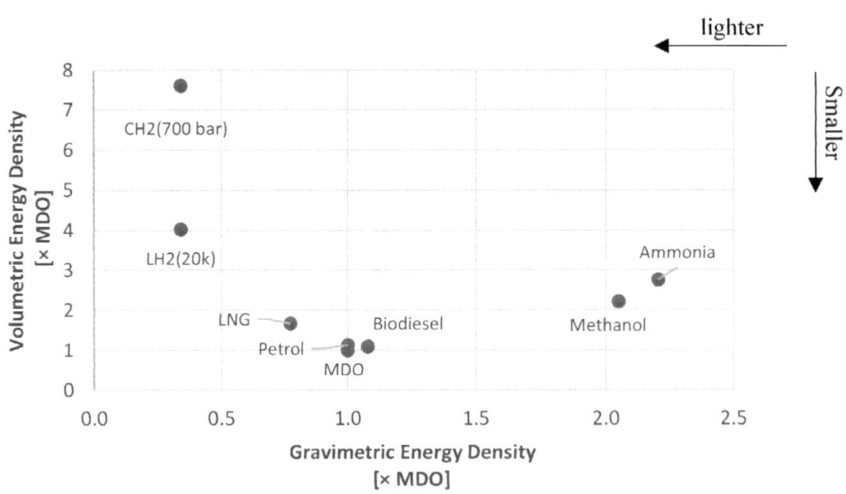

Figure 2: Gravimetric vs. volumetric energy density of fuels in comparison to MDO excluding the onboard storage system

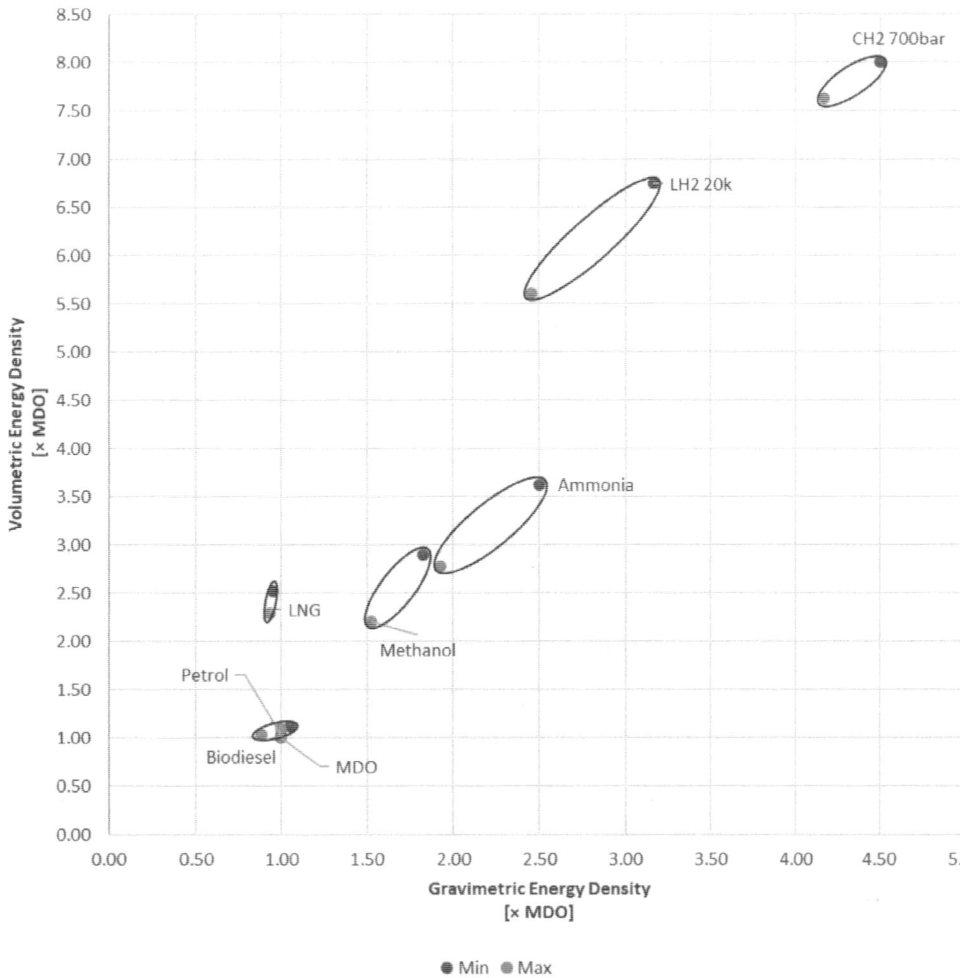

Figure 3: Gravimetric vs. volumetric energy density of fuels in comparison to MDO including the onboard storage system [16-21]

A battery-FC hybrid system comprises several components, including batteries, fuel cell stack, storage tanks, an electric motor propulsion system, fuel cell converters, and various handling devices like pumps, evaporators, air blowers, and filters. The total volume of the fuel cell propulsion system equipment for a small craft is assumed to be comparable to that of a diesel engine system [22] and therefore is not included in the current study's calculations. However, to accurately consider the volume and weight requirements for propulsion equipment, a detailed study focusing on machinery requirements based on power demand is necessary. The storage assumptions in this report are based on small storage tanks typically holding less than 300 litres of fuel, except for LNG, which has been assessed for larger vessels.

In Table 3, the specific fuel consumption (SFC) of alternative fuels is presented (based on their LCVs presented in Table 2) when they are used to power fuel cells, while MDO and biodiesel are used to power ICEs. It is generally recognized that fuel cells are more energy efficient than ICEs. For the purpose of this analysis, an assumption of 50% energy efficiency for fuel cells and 40% energy efficiency for ICEs has been made.

The results in Table 3 indicate a smaller difference in specific fuel consumption between the alternative fuels and MDO. Figure 4 represents the visual results of the analysis, highlighting the relative specific fuel consumption of the alternative fuels with MDO serving as the baseline for comparison. It is notable that values greater than 1 indicate higher specific fuel consumption compared to MDO.

Table 3. Specific Fuel Consumption of various fuels powering FCs and ICEs

Fuel-power unit	CH2- FC	LH2-FC	NH3-FC	MeOH-FC	MDO-ICE	Biodiesel-ICE
Efficiency (%)	50	50	50	50	40	40
SFC (m^3/MWh)	1.56	0.83	0.57	0.46	0.26	0.28
SFC (kg/MWh)	60	60	387	360	220	237

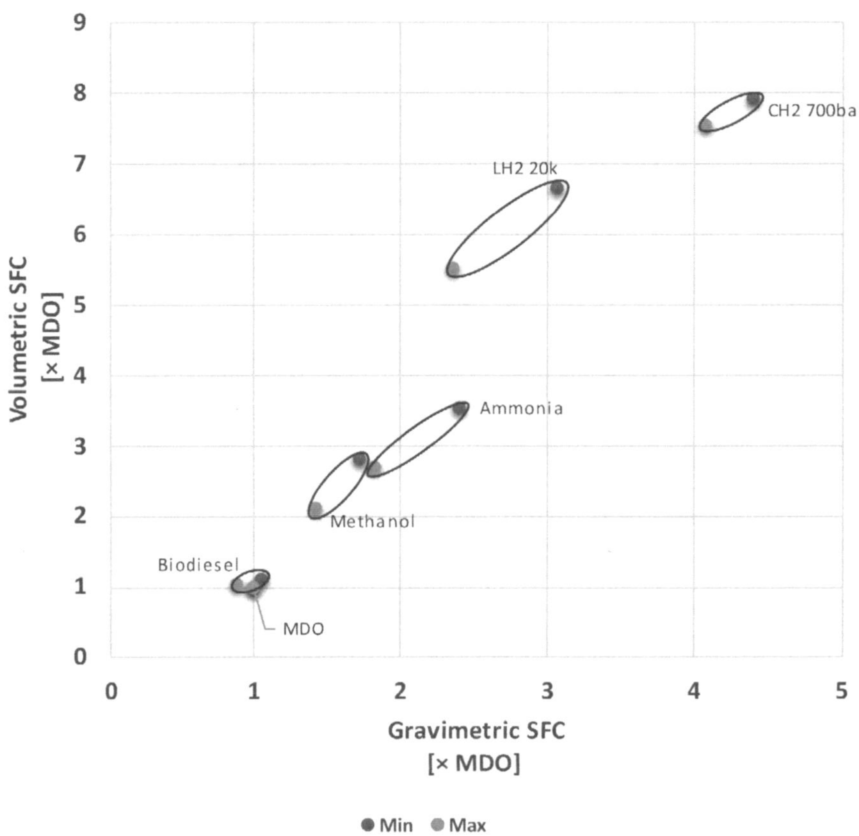

Figure 4: Gravimetric vs. volumetric specific fuel consumption with the assumption of Biodiesel and MDO powering ICEs and the alternative fuels powering FCs with 10% higher energy efficiency

4. SAFETY

The safety of fuel cells relies heavily on the safety features of the fuel itself, including properties such as density, flashpoint, auto-ignition temperature, toxicity, and other relevant factors. Some safety precautions may involve the use of gas-tight enclosures for fuel pipes and stacks to prevent leaks, regular monitoring for signs of leakage, designated vents to safely release any leaked fuel, and the installation of emergency shutdown systems to promptly respond to safety incidents or malfunctions. It is important to provide adequate training for new staff members regarding emerging risks and the maintenance of new technologies. However, due to the limited training time available for RNLI volunteers and their shorter time spent with the vessel compared to full-time crew members, it is crucial that systems are designed and integrated into the vessels and stations with a focus on ensuring the safety and effectiveness of the volunteers in their missions.

The greatest risk associated with hydrogen is the potential for fire and explosion due to hydrogen overflow or fuel cell control failure. The use of hydrogen can pose a higher risk for cargo vessels that transport dangerous goods. A previous study investigated the risk assessment of a passenger ferry with a hydrogen storage capacity of 450 kg over a 113-nm voyage in Norway during operation and while moored in the harbour, in accordance with the requirements of the IGF-code (international code of Safety for Ships using gases or other low flashpoint fuels). The results indicated a limited estimated risk associated with hydrogen systems, with less than 0.01 fatalities per 109 passenger km [23].

Methanol, with its low flashpoint of 11°C, increases the risk of fire and burns at low temperatures with an invisible flame and no smoke. Methanol tanks can be made from stainless or carbon steel with corrosion protection coatings. Essential safety measures for the use of methanol storage systems include venting, spill containment, vapour and fire detection, and crew training. Methanol's safety codes are under development which is similar to the IGF code for LNG. Fuel cells for methanol have been demonstrated onboard vessels. However, TRL 9 is expected around 2030 [24].

Ammonia is considered the preferred carbon-neutral fuel based on source-to-tank costs. However, its toxicity raises safety concerns, particularly for passenger vessels – as this study focuses on lifeboats, the combination of crew and rescued casualties emphasises the importance of considering potential harm caused by ammonia escape. Ammonia is toxic at low concentrations and poses a hazard to personnel. Nevertheless, a 5ppm odour can be noticeable by people. Deluge and detox systems are required, and an emergency extraction and independent ventilation system are essential.

Ammonia can be corrosive in the presence of moisture and reacts with acids and halogens. The use of stainless steel can protect materials from corrosion. Selecting appropriate materials for tanks, pipes, and structural components is essential, (similar to LPG carriers' safety standards).

Liquid ammonia can either be cooled down to -33°C or pressurised above vapour pressure. Storage of ammonia in solid metal ammines e.g. $Mg(NH_3)_6CL_2$, could provide additional safety by eliminating the risks associated with anhydrous ammonia [25]. Ammonia vapour is not considered a fire hazard due to its high ignition temperature, but concentrations of 15-28% can lead to an explosion. However, even a 0.5% vapour concentration can cause eye and skin irritation within a 20-minute exposure. Fortunately, ammonia vapour has a significantly lower density compared to air, causing it to disperse upwards and eventually dissipate at higher levels [6].

The main hazards associated with ammonia as a fuel for fuel cells include its toxicity, corrosiveness, and the potential release of liquid ammonia during bunkering operations. Bunkering of ammonia involves the transfer of both liquid and vapour and a spill can result in environmental damage. Proper crew training and handling procedures for ammonia are necessary. Lessons learned from its use as a fertiliser and chemical compound can be applied to maintenance, inspection, and ventilation procedures. It is worth noting that certain regions may consider banning the bunkering of ammonia due to safety concerns.

A research study has examined the safe use of ammonia fuel cells in the marine industry [6]. Currently, ammonia-powered vessels are not widely used. However, ammonia is already employed in other applications, and there are reports of risk assessment models for land-based facilities that estimate leak frequency and assess the physical effects on individuals. Existing shipping regulations related to the safe use of gases, such as the IGF regulations, need to be revised to provide specific guidelines for ammonia as a marine fuel.

5. ECONOMICS OF ALTERNATIVE FUELS

The cost projections for alternative fuels are subject to various factors, including market dynamics, governmental policies, technological developments, and the specific context of fuel utilisation. Hydrogen, as a core component of alternative fuels, plays a significant role in this field. The cost of hydrogen production can vary depending on the method used. Hydrogen produced from fossil fuels, known as grey hydrogen, is currently the most cost-effective option. However, it generates carbon emissions. On the other hand, green hydrogen, which is produced from renewable sources through electrolysis, is environmentally friendly but currently more expensive. It is expected that the cost of green hydrogen will become more competitive with grey hydrogen by 2030 as the technology develops with the market dynamics.

E-methanol and e-ammonia are anticipated to become more cost-effective with discount rates of 12% per year, reaching final targets of 24 GBP/GJ for e-methanol and approximately 18 GBP/GJ for e-ammonia by 2050 [24, 26]. These cost projections include the cost of raw materials, electricity, logistics, and OpEx excluding feedstock. It's notable that the cost of liquefying and transporting CO_2 is not considered in this study.

Table 4 presents a cost comparison of alternative fuels when used by fuel cells with MDO when used in ICE, specifically for an E-class lifeboat operating in the River Thames. The annual fuel consumption of the E-class lifeboat is assumed to be 10,000 litres based on the weekly consumption approximations. The calculations do not consider the effects of inflation and carbon tax. Additionally, the assumption is made that fuel cells have 10% higher energy efficiency than conventional systems.

The production of biofuels is expensive and has limited production volume since it may compete directly or indirectly with the food industry. Biofuels are not scalable solutions but could be economic for the aviation industry due to its higher energy demand and incompatibility with other fuels.

Table 4. Cost projection of alternative fuels for E-class boat between 2025-2050

Fuel	Price prediction in 2025 (GBP/GJ)	Price prediction in 2045 (GBP/GJ)	Total CO_2 generation in 25 years (tonne)	Average cost per tonne CO2 reduction until 2050 (GBP)	Cost 2025-2050 (x MDO)
MDO	11.2	18.4	776	0	1.0
E-methanol	48.8	28	0	215	2.3
Bio-methanol	20.8	19.2	233	31	1.2
Blue ammonia	20.8	18.4	581	26	1.2
E-ammonia	39.2	22.4	0	140	1.8
H_2	32.8	12.8	0	37	1.2
LH_2	37.6	17.6	0	84	1.5

6. ENVIRONMENTAL LIFE-CYCLE ASSESSMENT

There are alternative production pathways for non-fossil fuels, including biofuels and electrofuels, resulting in different well-to-wake emissions. Table 4 provides a comparison of well-to-tank and tank-to-wake emissions of different fuels in relation to MDO. "Well-to-tank" refers to the emissions associated with fuel production and transportation, while "tank-to-wake" emissions are generated during fuel combustion.

Hydrogen can be produced through Steam Methane Reforming (SMR) using natural gas which can be coupled with Carbon Capture and Storage (CCS), water electrolysis which can be powered by renewable energy sources, and gasification of coal or biomass which can also feature CCS. Hydrogen is considered a zero-emission alternative since the only emission produced is water. However, there are GHG emissions are generated during the production of hydrogen. Currently, most hydrogen production relies on natural gas reforming, which is a CO2-generating process. Carbon-neutral production methods are still in the early stages.

Methanol is a sulphur-free alcohol fuel which can be produced through several methods. One common approach is steam reforming, where natural gas is subjected to high temperatures and steam to convert methane into synthetic gas, which can then be further processed to produce methanol. Another method involves gasification, which can be applied to oil, coal, or biomass. This process entails heating these materials in the presence of oxygen to produce synthetic gas, which can be converted into methanol. When methanol is produced from natural gas, CO2 emissions can be equivalent to or even higher than those from fossil fuels.

Bio-methanol can be made from biomass and heat steam to produce bio-derived substitute fuel. Bio-methanol could reduce emissions by 70-100% compared to LSFO when biomass is used. The risk in the implementation of e-methanol is its low supply which is expected to reach 2% of fleet needs in 2035 [24].

E-methanol can be produced through the combination of hydrogen with a carbon source using renewable energy sources. This can be achieved through two processes: point source and Direct Air Capture (DAC). DAC methanol is considered a more sustainable option but is more expensive to produce. As point sources become scarce, the cost of e-methanol will depend on the cost of CO2 from direct capture.

Ammonia, as the most abundant globally produced chemical, is mostly produced by combining nitrogen with hydrogen through the Haber-Bosch process, which requires electricity and uses hydrogen and nitrogen as feedstocks. Nitrogen is produced from air through air separation, while hydrogen can be produced from natural gas, biomass, or through electrolysis using electricity. Water and pure nitrogen are the only emissions of ammonia and therefore the only pollutant it can generate is nitrogen oxide. Employing after-treatment systems to tackle this problem is essential when ammonia is used in conventional engines. Blue ammonia is produced through conventional methane reforming combined with CO_2 capture, while e-ammonia is produced through the electrolysis of water using renewable sources of electricity.

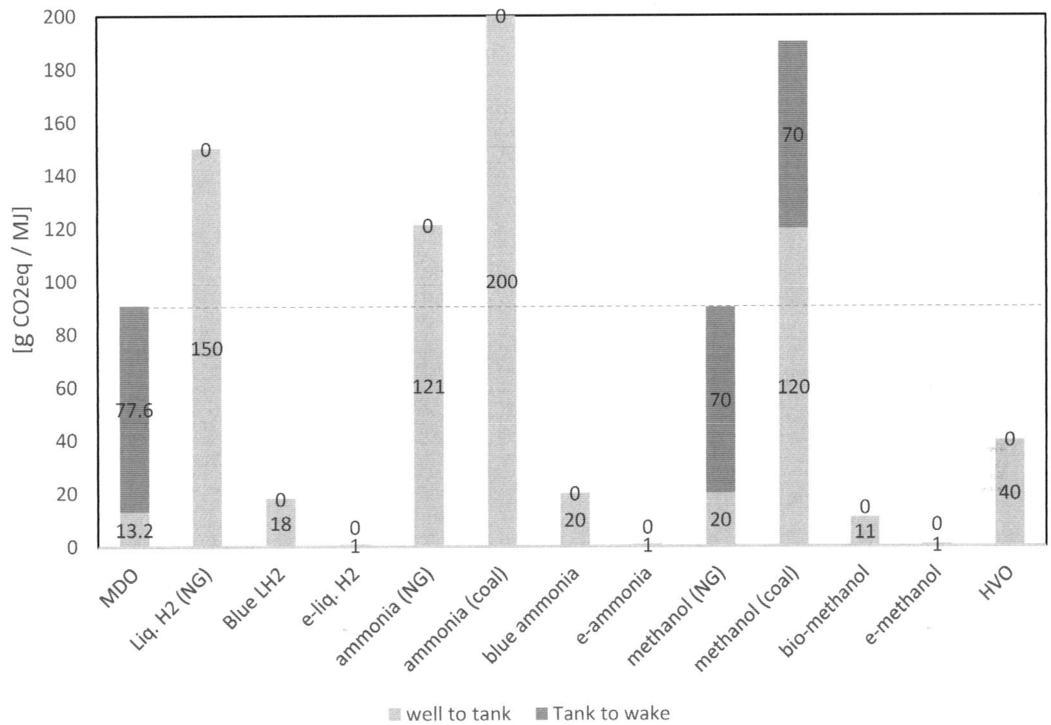

Figure 5: GHG emission of alternative fuels [24, 26-29]

7. INFRASTRUCTURE AVAILABILITY OF FUELS IN THE UK AND IRELAND

The UK's SMR technology companies are among the global emerging leaders [30], however, the country's strength is electrolysing technology is relatively underdeveloped. ITM Power, Johnson Matthey and Ceres Power have already started low and zero-carbon hydrogen technologies and are scaling up the production of green hydrogen.

The UK has established a strong technology in methanol synthesis with the presence of world-leading companies like Johnson Matthey and Amec Foster Wheeler. Furthermore, the UK's extensive onshore and offshore wind technology infrastructure provides a significant opportunity to enhance the production of e-methanol which relies on renewable energy sources. However, it is worth noting that a significant portion of the growth in the methanol market for the UK is happening outside of the country.

Johnson Matthey in the UK holds a leading position in Haber-Bosch technology. Moreover, Siemens UK has started a green ammonia plant in Harwell, Oxfordshire, powered by renewable energy sources. While the UK possesses strength in ammonia production technologies that enable low to medium-volume production, it faces challenges in maintaining competitiveness at high volumes. Factors such as high labour and input costs can impact the cost-effectiveness of large-scale ammonia production in the country. Therefore, access to international markets becomes crucial for the UK to sustain its share in the global ammonia market [31].

Figure 6 illustrates the location of some of the existing and developing infrastructure for alternative fuels in the UK and Ireland. Among these fuels, hydrogen occupies a significant portion as it serves as a fundamental component for other fuels and also functions as a fuel itself. It is important to note that while the illustration highlights various types of infrastructure, the focus on biofuels primarily revolves around the production of sustainable aviation fuel (SAF) which has not explicitly mentioned in the figure.

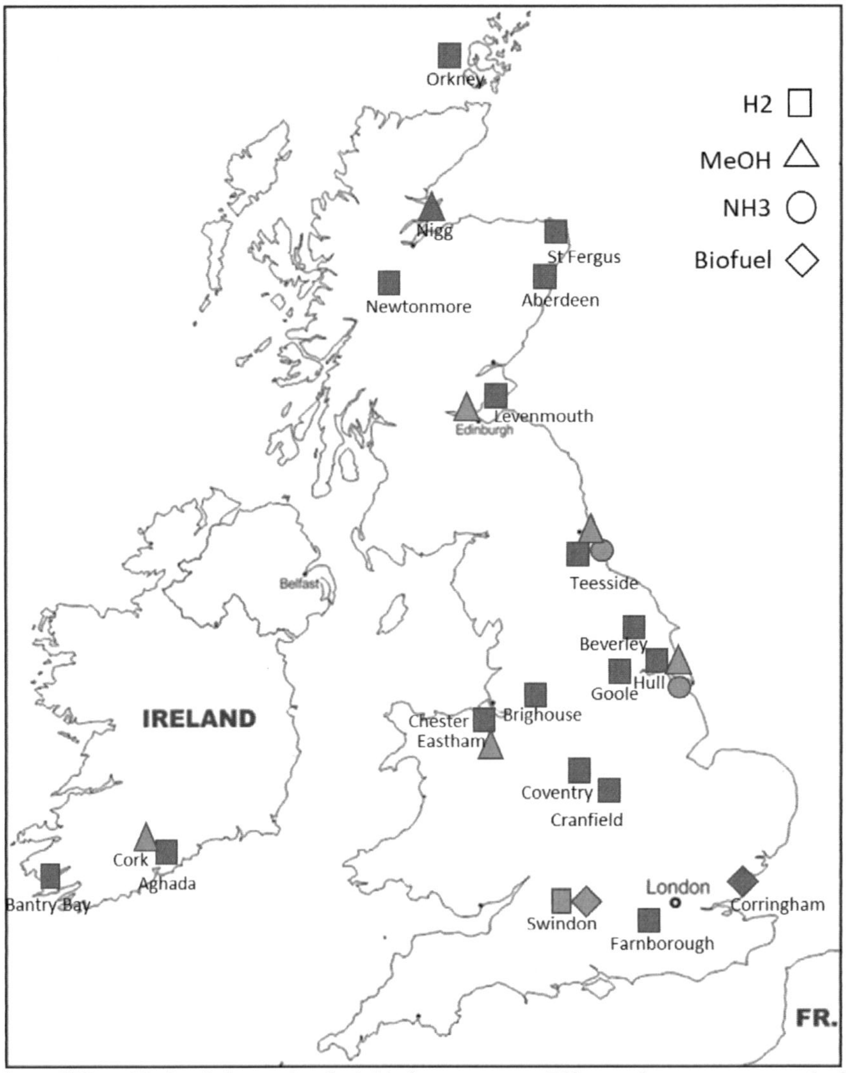

Figure 6: Some UK and Ireland alternative fuels' infrastructure – different shapes represent alt. fuels – with green indicating operational status and blue indicating ongoing projects

Figure 7 illustrates the relative performance of alternative fuels based on the criteria outlined in this study. E-fuels' key strength is their potential to reach carbon neutrality. However, their current infrastructural limitations lead to high production costs which is considered a significant barrier to their widespread adoption. Updates to safety regulations will play a significant role in enabling their use in the maritime industry.

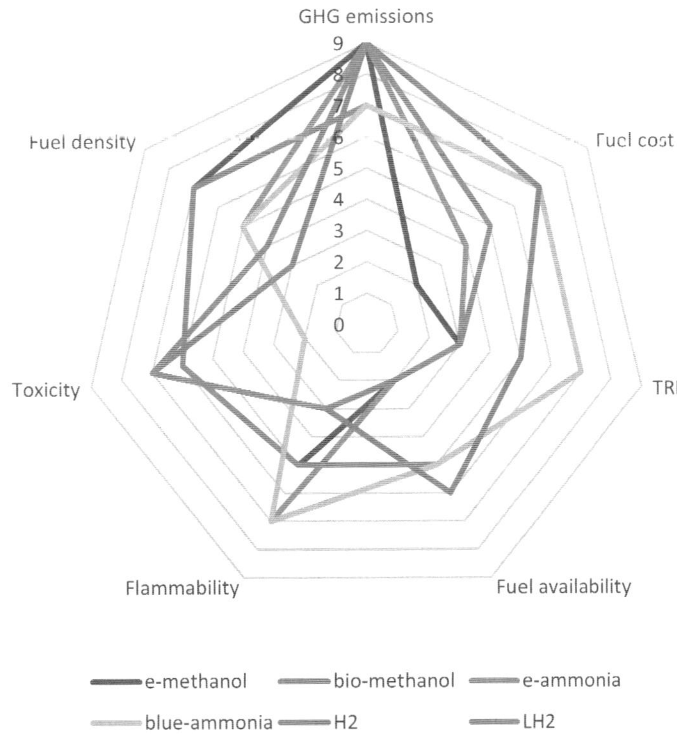

Figure 7: Relative performance of alternative fuels – higher numbers indicate better performance

8. CONCLUSIONS

The review focused on comparing low and zero-carbon fuels to be used by fuel cells for the RNLI fleet. High relative specific energy of fuel cells in combination with higher specific power of batteries can potentially provide longer endurance with quicker response to high-power demands compared to using them individually. However, further assessment is required to evaluate the impact of actual system design including space constraints, weight effect, safety considerations, etc on the individual boats in the RNLI fleet.

PEMFCs (mainly fuelled by hydrogen) are among the most developed types of fuel cells, followed by methanol-fed fuel cells, and ammonia-fed fuel cells. However, the main barrier to replacing ICEs with fuel cells is the cost. Until fuel cells become more cost competitive with ICEs, the considerably higher CapEx associated with implementation remains a significant deterrent for vessel owners.

Biofuels may not be scalable solutions for the marine industry due to the greater demand in the aviation sector. Nevertheless, considering their availability in the UK, they can serve as suitable options during the transitional period, particularly for small craft such as lifeboats, which cover a smaller proportion of the whole energy demand in the maritime industry. Furthermore, hybrid ICE-battery systems can play a substantial role until higher maturity and cost competitiveness of fuel cells.

Volume and weight efficiency of alternative fuels along with their storage space, were studied in comparison to ICEs fuelled by MDO. Methanol fuel storage, despite requiring on average 2.5 times the size of MDO storage and being on average 1.5 times heavier than MDO fuel storage system, demonstrated the best performance among the e-fuels.

An economic comparison of fuels cost was conducted, with MDO as the baseline fuel, for an E-class lifeboat with an average annual fuel consumption of 10,000 litres, covering the period from 2025 to 2050. Compressed hydrogen, bio-methanol, and blue ammonia showed the most cost-effective outcomes. On the other hand, e-methanol, e-ammonia, and liquid hydrogen are the most expensive options in the given order. Ammonia as the cheapest e-fuel, holds a significant potential role in the transition to alternative fuels. However, the extent of ammonia's role could be limited by a rapid decline in renewable electricity cost which can make other e-fuels more cost-effective.

A comparison of environmental emissions from fuels was conducted, considering both the well-to-tank and tank-to-wake perspectives. It was found that methanol, when produced from natural gas, can result in similar GHG emissions as MDO. Ammonia produced from natural gas and coal, as well as methanol produced from coal, may even generate higher GHG emissions than MDO throughout their lifecycle. Hence, the development and use of renewable sources in the production process of alternative fuels are crucial factors in this endeavour.

Safety regulations for methanol, being the preferred choice for larger vessels with high energy density requirements, have been more extensively developed. On the other hand, safety codes for the onboard use of ammonia are still progressing due to its highly corrosive and toxic nature, requiring additional safety precautions. This is particularly important for RNLI lifeboats which have a compact design and the potential for carrying a high density of people onboard. In the case of hydrogen, which is a highly explosive fuel, the development of safety regulations has been slower due to its low energy density and limited adoption in large vessels.

The UK, with its world-leading companies in e-fuel production technology and abundant onshore and offshore renewable power generation sources, holds significant potential for the future production of e-fuels. However, challenges may arise in maintaining global competitiveness, particularly at high production volumes. Efforts will be needed to address these challenges and ensure that the UK remains competitive in the evolving market for e-fuels.

Based on preliminary options assessments, there appears to be potential for fuel cell hybrid arrangements as a future power source for some maritime SAR craft. However, it should be noted that this study does not conclude that this is the preferred option for the RNLI vessels and there is ongoing research in this field to thoroughly explore and evaluate the available options, as well as their impact on both existing and potential new designs of SAR vessels.

9. ACKNOWLEDGEMENTS

The research reported in this paper has been partially funded by the RNLI. The authors wish to acknowledge the contributions and support provided by the RNLI charity, as well as Neil Chaplin, Spencer Johns, and Victoria Limbrick from the RNLI for their support in undertaking this research and preparing the paper.

10. REFERENCES

1. JOUNG, T H., et al., 'The IMO initial strategy for reducing Greenhouse Gas (GHG) emissions, and its follow- p actions towards 2050'. *Journal of International Maritime Safety, Environmental Affairs, and Shipping,* 2020. **4**(1): p. 1-7.

2. IMO. Intersessional Working Group on Reduction of GHG Emissions from Ships (ISWG-GHG 14), 20-24 March 2023. 2023; Available from: https://www.imo.org/en/MediaCentre/MeetingSummaries/Pages/Intersessional-Working-Group-on-Reduction-of-GHG-Emissions-from-Ships-(ISWG-GHG-14),-20-24-March-2023.aspx.

3. RNLI.. 'Environmental sustainability at the RNLI. 2023' Available from: https://rnli.org/about-us/sustainability/environmental.

4. INAL, O B., and DENIZ, C., 'Assessment of fuel cell types for ships: Based on multi-criteria decision analysis'. *Journal of Cleaner Production*, 2020. **265**: p. 121734.

5. XING, H., et al., 'Fuel Cell Power Systems for Maritime Applications: Progress and Perspectives'. Sustainability 2021, 13, 1213. 2021, s Note: MDPI stays neutral with regard to jurisdictional claims in published ….

6. CHELIOTIS, M., et al., 'Review on the safe use of ammonia fuel cells in the maritime industry'. *Energies*, 2021. **14**(11): p. 3023.

7. DE-TROYA, J J., et al., 'Analysing the possibilities of using fuel cells in ships'. *International Journal of Hydrogen Energy*, 2016. **41**(4): p. 2853-2866.

8. PRATT, J W., and KLEBANOFF, L E., 'Feasibility of the SF-BREEZE: a zero-emission, hydrogen fuel cell, high-speed passenger ferry'. 2016.

9. LANGFELDT, L., 'Maritime Fuel Cell Applications'. 2018.

10. OPITZ, A., et al., 'Can Li-Ion batteries be the panacea for automotive applications?', *Renewable and Sustainable Energy Reviews,* 2017. **68**: p. 685-692.

11. CHICHE, A., et al., 'Feasibility and impact of a Swedish fuel cell-powered rescue boat'. *Ocean Engineering*, 2021. **234**.

12. Torqeedo. Electric mobility on the water. 2022; Available from: https://www.torqeedo.com/en/dealers/commercial-use.html.

13. WINTER, M., and BRODD, R J., 'What are batteries, fuel cells, and supercapacitors?', *Chemical reviews*, 2004. **104**(10): p. 4245-4270.

14. Calgary, U.o. Energy density. [cited 2023 April]; Available from: https://energyeducation.ca/encyclopedia/Energy_density.

15. ABE, J O., et al., 'Hydrogen energy, economy and storage: Review and recommendation', *International journal of hydrogen energy*, 2019. **44**(29): p. 15072-15086.

16. FORETICH, A., et al., 'Challenges and opportunities for alternative fuels in the maritime sector', *Maritime Transport Research*, 2021. **2**: p. 100033.

17. ACEVES, S M., et al., 'High-density automotive hydrogen storage with cryogenic capable pressure vessels', *International Journal of Hydrogen Energy*, 2010. **35**(3): p. 1219-1226.

18. VAN VELDHUIZEN, B., et al., 'Comparative Analysis Of Alternative Fuels For Marine Sofc Systems', 2022.

19. VAN BIERT, L., et al., 'A review of fuel cell systems for maritime applications', *Journal of Power Sources*, 2016. **327**: p. 345-364.

20. GEERTSMA, R., and KRIJGSMAN, M., 'Alternative fuels and power systems to reduce environmental impact of support vessels', *Delft University of Technology*, 2019.

21. Maritime, D.G.A., *Comparison of Alternative Marine Fuels*. 2019.

22. YE, M., et al., 'System-level comparison of ammonia, compressed and liquid hydrogen as fuels for polymer electrolyte fuel cell powered shipping', *International Journal of Hydrogen Energy*, 2022. **47**(13): p. 8565-8584.

23. AARSKOG, F G., et al., 'Concept risk assessment of a hydrogen driven high speed passenger ferry', *International Journal of Hydrogen Energy*, 2020. **45**(2): p. 1359-1372.

24. Centre, M.M.-K.M., *Methanol as a marine fuel*. 2021.

25. ZHAO, Y., et al., 'An efficient direct ammonia fuel cell for affordable carbon-neutral transportation', *Joule*, 2019. **3**(10): p. 2472-2484.

26. Centre, M.M.-K.M., *Ammonia as a marine fuel*. 2021.

27. Centre, M.M.-K.M., *We show the world it is possible*, in *Documentation and assumptions for NavigaTE 1.0*. 2022.

28. LINDSTAD, E., et al., 'Reduction of maritime GHG emissions and the potential role of E-fuels', *Transportation Research Part D-Transport and Environment*, 2021. **101**.

29. XING, H., et al., 'Alternative fuel options for low carbon maritime transportation: Pathways to 2050', *Journal of Cleaner Production*, 2021. **297**: p. 126651.

30. Secretary of State for Business, E.I.S., *UK Hydrogen Strategy*. 2021.

31. Economics, F., *reducing the maritime sector's contribution to climate change and air pollution*. 2019.

11. AUTHORS BIOGRAPHY

Nickan Ghahramani holds the current position of PhD researcher at Newcastle University with a focus on low and zero-carbon fuels in the context of RNLI vessels. His previous experience includes lecturing for various engineering modules and has practical experience as a marine engineer, having sailed on ocean-going vessels.

Dr Alan J Murphy, BEng, PhD, CEng, CMarEng, FIMarEST, FRINA, FHEA, is a Reader (Associate Professor) in Maritime Engineering at Newcastle University. He was formally a sea-going Marine Engineer. For more than a decade he has led a large portfolio of UK government, European Union and Industry-funded research projects, many broadly on the theme of sustainable shipping through the reduction of airborne pollution and improved energy efficiency of marine machinery systems.

Dr Simon Benson is a Senior Lecturer in Naval Architecture at Newcastle University. He is responsible for teaching and research in marine hydrodynamics and structures. His previous experience is in industry working for BMT Fluid Mechanics and IHC Engineering Business before completing a PhD investigating the structural performance of aluminium high-speed ships. His PhD and post-doctoral work were sponsored by the Office of Naval Research.

Prof Richard Birmingham is Professor Emeritus of Small Craft Design at Newcastle University where he has been a member of the academic team for over twenty-five years, including seven as Head of School. He previously served as the president of The Royal Institution of Naval Architects. He has worked on collaborative research projects with the RNLI for over 20 years and he is currently a member of the RNLI's Technical Advisory Panel. His research interests include design for sustainability in its widest sense and in particular in the context of artisanal fishing boats in the developing world.

Neil Chaplin holds the current position of Head of Asset Management at the Royal National Lifeboat Institution where he has worked for over 2 decades. He is a member of the senior leadership team in the Engineering and Supply department of the RNLI. Responsible for the development and implementation of asset strategies as well as holding the delegated role of RNLI chief engineer, responsible for technical standards of RNLI assets. Prior to joining the RNLI he was employed by the UK Ministry of Defence procurement with experience in survey vessels and Royal Fleet Auxiliary vessels. His previous experience at the RNLI includes project manager for the RNLI Tamar class ALB design and build, Principal Naval Architect and a Staff Officer in RNLI operations.

Spencer Johns BEng, IEng, MIMechE is Principal Engineer (Marine / Mechanical) at the Royal National Lifeboat Institution. Responsible for the design, development and thorough life support of the main machinery and auxiliary systems onboard all classes of lifeboats. He has worked extensively on propulsion and driveline systems, since joining the RNLI in 2000 and was project manager for the RNLI Severn class ALB engine upgrade programme.

Victoria Limbrick worked as Carbon & Energy Manager for the RNLI at the time of writing, with responsibility for developing decarbonisation strategies for RNLI assets and activities, introducing alternative low/zero carbon technologies to the organisation and monitoring and reporting on carbon emissions, energy and fuel usage. Victoria has worked in energy efficiency and carbon reduction since 2005 starting in the construction industry and then for RNLI since 2015, now working as an Energy Manager for Balfour Beatty.

DEVELOPING A MOTIONS ANALYSIS PLATFORM TO VALIDATE THE BENEFIT OF COURSE AND SPEED CHANGES TO IMPROVE PASSENGER COMFORT AND EQUIPMENT INTEGRITY

S McCance, and **Dr T Newman**, Naval Design Partnering, United Kingdom

SUMMARY

NDP research has quantified the very apparent benefits that a modest reduction in speed in a seaway can achieve with regards to reducing exposure of equipment and personnel to extreme and aggregate shock and vibration events in differing sea states. This has been shown in simulator trials using the Fast Small Ship Simulator (FSSS) operated by MARIN where the behaviour of several experienced coxswains was evaluated with the advantage of controlled conditions when compared with previous sea trials. Here identical conditions and influences were repeated with the objective to quantify the impact of both sustained transit speed and course on shock and acceleration.

In support of validating the simulator trial, the NDP has been tackling the significant challenge of reliably quantifying sea state in small craft at speed using its custom developed motion sensor platform (named WaveReader). The representation of sea states is understood statistically and hence a breadth of significant wave heights can be predicted from analysis of wave motions. This is a proven technology in stationary wave buoys however is less robust in refined hull forms and whilst underway.

WaveReader has been developed to validate a simulator study in a real-world scenario alongside moving the research capability from simple data collection, followed by post-processing and analysis, forward to a system capable of real-time processing of data streams and rapid calculation allowing for predictive outputs to be utilised onboard by crew and coxswains. Whilst the WaveReader equipment has presented a significant leap forward in the equipment set available for NDP motions analysis research its focus and intent to provide a helm aide for high-speed operations is under active development. Isolated trials can be limited in their ability to provide robust conclusions and hence the long-term testing of the WaveReader device will allow the simulator outcomes to be proven in a real-world scenario.

The key conclusion from the simulator research was quantifying the impact that controllable changes to operations have on the aggregate and instantaneous scale of exposure to boat motions. Exposure to boat motions has been quantified in TTLV and VDV and it has been shown that a speed reduction of some 5% from top speed can equate to a reduction in VDV of some 20%, this relationship can be seen in Figure 5, confirming that planners can significantly reduce WBV risk to crew with a modest reduction in average speed.

What has however been apparent throughout a wide range of trials is that with any mitigations towards exposure to WBV effects is that mitigations should be used to reduce the impact on personnel and equipment and to increase efficiencies rather than providing safeguard for increasing overall operational speed.

NOMENCLATURE

Abbreviations:

DFT	Discrete Fourier Transform
FFT	Fast Fourier Transform
FSSS	Fast Small Ship Simulator
KNRM	The Royal Netherlands Sea Rescue Institution (Koninklijke Nederlandse Redding Maatschappij)
MARIN	Maritime Research Institute Netherlands
MOD	Ministry of Defence, United Kingdom
NDP	Naval Design Partnering
RNLI	Royal National Lifeboat Institution
TTLV	Time to Limit Value
WBV	Whole Body Vibration
VDV	Vibration Dose Value

Symbols:

$A(t)_W$	Weighted acceleration time history (m/s2)
H_S	Significant Wave Height (m)
m_n	n^{th} moment of the power spectrum
S	Power Spectrum
T	Duration over which the VDV has been calculated (s)
T_{VDV}	Time taken to reach the VDV threshold (s)
T_Z	Zero Crossing Period
$VDV_{threshold}$	Daily limit value ()
ω	Angular Frequency (rad/s)

1. INTRODUCTION

The representation of sea states is understood statistically and hence a breadth of significant wave heights can be predicted from analysis of wave motions. The NDP has been tackling the significant challenge of reliably quantifying sea state at speed using its custom developed WaveReader motion sensor platform. WaveReader has been developed to validate a simulator study in a real-world scenario alongside moving the research capability from simple data collection, followed by post-processing and analysis, forward to a system capable of real-time processing of data streams and rapid calculation allowing for predictive outputs to be utilised onboard by crew and coxswains.

Research and development, supporting the ongoing Whole Body Vibration (WBV) mitigations undertaken by UK MOD for its small craft operations, has quantified the very apparent benefits that a modest reduction in speed in a seaway can achieve with regards to reducing exposure of equipment and personnel to extreme and aggregate shock and vibration events in differing sea states. This has been shown in simulator trials using the Fast Small Ship Simulator (FSSS) operated by MARIN for the Dutch navy where the behaviour of several experienced coxswains from MOD, RNLI and KNRM was evaluated with the advantage of controlled conditions when compared with previous sea trials. Here identical conditions and influences could be repeated where the main study objective was to quantify the impact of sustained transit speed and course on WBV and shock using the simulator and from which several trends have been quantified.

This paper presents details of the simulator trials which has quantified the relationships between sea state, shock exposure and helm control and subsequently documents the progress of the NDPs WaveReader platform developed for monitoring sea state and tracking coxswain behaviour in order to validate the simulator findings.

2. SIMULATOR TRIALS

2.1 BACKGROUND

A large body of research has been undertaken by maritime organisations to characterise shock and associated Whole-Body Vibration (WBV) health impacts. WBV exposure of RHIB occupants is exceedingly difficult to predict due to the complex nature of the sea-boat-human interactions.

From a procurement perspective, boat requirements are driven by desired capability and the associated performance of this capability. On paper, it is straightforward to set a requirement to achieve a particular transit in a given time however in reality there are two variables of which the impact is very hard to quantify: the sea state, and the coxswain's ability to navigate said sea state. Sea state is managed in requirements by specifying a required speed in a chosen sea state, but it is difficult to prove how a boat will perform when these two variables strongly interact, and what impact that will have on shock exposure.

Whilst sea trials and research can provide insight into the problem and data for analysis, the infinite variability of a seaway and the relative behaviour of a skipper in reaction to this is harder to quantify. Previous NDP research has looked at tracking coxswain behaviour by way of both video analysis and of utilising electronics customised to individual craft. This has none the less been limited to isolated trials where cost and availability and other factors have prevented side by side trials to be carried out or extensive data sets to be acquired. Following the build of now two WaveReader equipment sets, further sea trials are planned for the coming year in which a side-by-side commercial trials and long-term testing is planned.

Minimal published research is available on tracking coxswain behaviour in rough seas, this likely due to the technical challenges associated with:
- sea trials – require a substantial number of variables to be recorded, including the boat control signals and sea conditions, which is difficult to do accurately/repeatably; and,
- simulator trials – few simulators exist which accurately re-create the boat physics and coxswain stimuli associated with operating in rough seas.

Use of the MARIN FSSS, as a calibrated simulator tool, has overcome the drawback associated with the integrity of simulator trials and has allowed a series of trials to be carried out in controlled conditions thus eliminating the variable element of the seaway and imposing a controlled condition on the participants with regards to the environment.

2.2 SIMULATOR STUDY

2.2.1 Outline

The MARIN FSSS was used to investigate the impact of coxswain behaviour on shock and vibration effects. Using a standard trials pattern Figure 1, coxswains were required to travel in six directions relative to the prevailing swell direction covering all points of sail, for a distance of 1.5km per leg. A series of successive trial runs were conducted represented a sequential increase in prevailing swell of sea state 2 through 4 based on the JONSWAP spectrum. Participants, with varying experience levels and from a selection of operators, were observed during the trial and debriefed following their participation.

2.2.2 Objectives

The objective of the overall study was to undertake coxswain trials to quantify impact sustained transits on shock, with capability to:

a. Analyse transit times for experienced and inexperienced coxswains in repeatable conditions for different scenarios
   i. full control, using both throttle and steering inputs
   ii. partial control, coxswain controlling throttle only
   iii. partial control, coxswain controlling steering only

b. Expose coxswain to a range of sea states in a controlled, controllable and recordable way

c. Determine relative importance of throttle and steering

d. Assess WBV exposure metrics and optimal behaviours

e. Relate the following cues for the coxswain to trends in how the subject responds, as a function of the conditions:
   i. when the throttle is used prior to the shock events.
   ii. how much the throttle is reduced by or increased.
   iii. how long the throttle is reduced or increased for, and.
   iv. steering usage, acknowledging that the throttle is more important according to literature.

2.2.3 Simulator Limitations

Whilst the simulator study approach provides the basis for replicating a controlled scenario there were a number of limitations and drawbacks including:

1. Coxswains may behave differently to how they behave on a real boat. The perception of risk by coxswains will be different to a real RHIB as there is little or no danger associated with exceeding the limits of the simulator, unlike in real world use cases. There will always be coxswains who drive too aggressively because it is considered more of a game.

2. Impacts 'hurt less' on a simulator. The peak accelerations are limited by the motion platform and are lower than output by the boat model. The limitation identified is the shortfall in peak acceleration between real impacts (of the order of 8g) and the limits of the motion base (approximately 2g). Simulator motion platforms with higher peak accelerations, however, are currently being developed.

3. Some experienced coxswains moderate their speed based on the impacts they feel and are therefore more likely to drive fast (unless instructed not to do so, as in this study). This could skew the results without sufficient pre-planning. At the development stage of FSSS, feedback was gathered from experienced

coxswains on whether the cues for wave slams are accurate. Visually, the cues were adjusted to provide a realistic experience to the coxswain based on their experience and feedback.

In terms of peak acceleration and body forces for a heavy landing, feedback was that perhaps 'knee pain would be higher in reality'. In many respects, this is a good sign that the visual cues were correct even if the forces/pain experienced were not. It is worth further highlighting that the aim of the study is to monitor how a good coxswain is able to reduce high magnitude impacts. Therefore, in theory, there shouldn't be a significant number of impacts excessively above 2g. The boat performance is still modelled at higher acceleration levels so the negative impact on transit speed should still be accurate even if the physical simulator motion response is not.

2.3 SIMULATOR RESULTS

2.3.1 Processing

Data and observations recorded during the simulator trial were subject to post processing and analysis to observe trends and outcomes. Data was spatially filtered prior to analysis to remove unwanted parts of the time histories, thus ensuring comparable transit segments were used as illustrated in Figure 2. This resulted in six separate legs for the different headings of approximately equal distance covered. Evidently, within a given spatial window, the distance could vary depending on the path taken by the coxswain through the waves.

The following data areas were removed:
1. Corners: where the heading changes significantly.
2. Initial acceleration and 'false starts': the boat getting up to speed at the run start and occasional short delays in starting a run where the boat could drift.
3. Overshoots: at the end of the run, some coxswains stopped the boat while others carried on.

The most common ways to describe shock and vibration experienced by a human occupant are the Root-Mean-Square (RMS) acceleration, the 8-hr normalised level (*A(8)*), and the Vibration Dose Value (*VDV*). Whilst the *A(8)* value is preferred for a majority of vibration applications, it is not well suited to evaluating the exposure to combined shock and vibration, as it is insensitive to one-off, high-magnitude events typically associated with high-speed craft operation. Acceleration time histories containing shock events are better represented by the VDV which is an effective dose-response relationship between the human body and the input acceleration.

2.3.2 Vibration Dose Values

The VDV of the acceleration time history is calculated as the integral of the fourth root of the weighted acceleration time history raised to the fourth power:

$$VDV = \left(\int A(t)_W^4 \, dt\right)^{\frac{1}{4}}$$

Where:
$A(t)_W$ is the weighted acceleration time history, with the exact weighting depending on direction of stimulus, point of input to the human body, and purpose of the vibration assessment. The integration is made with respect to time. For this analysis, the vertical vibration up into the human body is considered using the Wk weighting factor defined in ISO 2631-1 (1).

2.3.3 Time To Limit Value (TTLV)

The TTLV is back calculated from the overall VDV for the given period to generate the amount of time an activity can be carried out for before an occupant is above the exposure limit value. The following equation is used

$$T_{VDV} = T\left(\frac{VDV_{threshold}}{VDV}\right)^{\frac{1}{4}}$$

Where:
T_{VDV} is the time taken to reach the VDV threshold
$VDV_{threshold}$ is Daily limit value = $21\ ms - 1.75$
T is the duration over which the VDV has been calculated

2.4 RESULTS

Whilst coxswain experience would seem to enable a transit to be undertaken at a faster speed, this increased skill level does not appear to have a pronounced benefit to the accelerations experienced by the craft. The effect of reduction in speed however has been shown to be significant. The calculated VDV for the study participants, presented in order of experience can be seen in Figure 3 where VDV generally increases with the more experienced coxswain operators. This is presented in contrast with Figure 4 which presents the corresponding transit speeds between participants as a function of sea state.

The relationship of the reduction to speed can be seen in Figure 5 alongside the TTLV data for differing speeds and sea states in Figure 6. Notably, a speed reduction of some 5% from top speed equates a reduction in VDV of some 20%. This indicates that planners could significantly reduce WBV risk to crew with a modest reduction in average speed. Beyond 15% speed reduction, there is a case of diminishing returns such that a further 15% reduction only equates to an additional 10% reduction in VDV compared to 60% for the first 15% reduction. This relationship is based on simulator outputs and requires further validation through at sea trials for which the WaveReader platform has been developed.

Figure 7 shows data for VDVs at differing headings for sea state 4, while the transit times and hence speeds were similar for each heading, the overall VDV is dominated by the exposure during either the head or bow seas. This is as expected. When travelling at most other headings, the coxswain was observed to mitigate large impacts by navigating a path that does not involve such abrupt changes accelerations and slamming.

Time to Limit Value (TTLV) curves have been offered as an aid in operations planning however these are currently limited to the parameters of the simulator craft, a high-speed RIB, which offered a calibrated model. There would likely be some level of discrepancy between this and specific service boats, however the importance of taking into account the prevailing sea conditions into operations planning on WBV has been demonstrated.

Overall TTLV curves are considered non-conservative under certain sea conditions. Effective throttle control has been shown to be much more important than steering to reduce WBV. In a follow-up to this study, the intent has been to analyse data records with regard to throttle and steering responses, on a per impact basis, will be assessed to understand the trends and optimal behaviours of one coxswain relative to another. Data is available for this analysis in due course.

3. REALTIME MOTIONS ANALYSIS

3.1 OVERVIEW

The representation of sea states is understood statistically and hence a breadth of significant wave heights can be predicted from analysis of wave motions. The NDP has progressed the development of a robust and customisable motions analysis platform that can reliably be used to analyse motions inputs from small craft with algorithms for quantifying wave height, sea state and the capability of studying operator inputs. The hardware development had been driven by the need to validate the findings of the simulator study with the goal of providing a helm aide for safeguarding personnel and assets.

The platform, termed WaveReader, Figure 8, has been developed to provide features and benefits not found collectively in COTS equipment and has been developed to be reliable, robust and portable; such that it can be deployed with ease thus requiring minimal or no SQEP support for either installation, calibration or for background data collection. A key requirement to aid the ongoing development of the equipment during longer term or unsupported trials.

The key goal of WaveReader has been to provide a data analysis platform that can offer the capability of real-time data and calculation feedback such that with continuing the refinement of onboard algorithms, the need for post-processing of data can be eliminated and where desired, key outputs can be used by operators. The drive, alongside validating the findings from the simulator, has been to develop a helm aide to improve safety of personnel and equipment where an operator can be advised on the likelihood of severe impact events and other feedback such as suggested course and speed changes and motions assessments; provided as proactive rather than reactive predictions.

3.2 SEA STATE

3.2.1 Wave Height

The key parameters to characterise sea state are Significant Wave Height and Zero Crossing Period: The calculation of wave height had formed the first challenge for the development of WaveReader with the goal of reliably quantifying this output in each stationary (anchored or station holding), adrift and underway in both the displacement and planing regimes.

Wave height has been quantified from motions normalised to an earth fixed axis system. The earth-fixed system presents a difficulty in describing motion of a body itself however allows motion measurements to consider heave as a validated sensor output. This does not account for the geometric position of the sensor within a vessel thus body centred output must also be considered so allow a comparison and calculation from multiple positions. In assessing the combined output with a knowledge of the geometric positions of the sensor suite comprising of the base station IMU alongside a number of remote accelerometers, a rationalised heave measurement can be quantified.

The spectral analysis method was used to calculate sea state parameters from heave data. The heave signal is calculated over 100 seconds segments as shown in Figure 9. A Fourier Transform expresses the heave signal as a sum of periodic components of different frequencies. The Discrete Fourier Transform (DFT) is defined as follows and can be calculated quickly using a Fast Fourier Transform (FFT):

$$y[k] = \sum_{n=0}^{N-1} e^{-2\pi j \frac{kn}{N}} x[n]$$

The power spectrum of the heave signal is then:

$$S[k] = |y[k]|^2$$

The zeroth moment of the power spectrum can be interpreted as the variance of the sea surface elevation, and is calculated as:

$$m_0 = \int_0^{\infty} S[\omega]\, d\omega$$

where wave frequency $\omega = \frac{2\pi}{\omega}$.

The significant wave height is then:

$$H_S = 4\sqrt{m_0}$$

To calculate the average zero crossing period it is necessary to also calculate the second moment of the power spectrum:

$$m_2 = \int_0^{\infty} f^2 S[\omega]\, d\omega$$

The zero-crossing period is then:

$$T_Z = 2\pi \sqrt{\frac{m_0}{m_2}}$$

Lloyd (2) alongside Lewandowski (3) present methodologies for analysis of seas states and analysis of the behaviour of surface craft, from which the WaveReader calculations have been devised. Throughout the development of the WaveReader device there has been the requirement for simplicity in installation and use hence the calculations assume only the onboard device sensors are available and any external feeds are supplementary to the calculation; hence a low confidence prediction can be made with the unit in isolation. Where any calculation is refined through the additional data feeds the confidence can then be improved.

Analysis of roll accelerations has, with an adjustment for sensor position, shown a good correlation with both known sea states and for positioning the centre of flotation broadly aligned with the vessel centreline.

3.2.2 Wave Direction

Heave motions have been the principal focus of the WaveReader device to date given their direct impact on accelerations and motions experienced onboard. Wave direction is also of interest and whilst to date this has been left to a visual appraisal augmenting the observations of variation in severity of the heave accelerations, further work is planned to apply directional algorithms to the calculation set.

The technique in Steele et al. (4) appears to be widely used in commercial wave buoys. This method uses a Fourier decomposition to estimate the mean and principal directions as the first two components of the spectrum. It was further concluded by an investigation undertaken by the English Institute of Sport that it may further be possible to differentiate wave types by focusing on elemental parts of the spectrum.

Algorithms defining wave direction are notably less well developed in the current equipment and form part of the ongoing development planned in the current project funding. Furthermore, the challenge of not only quantifying wave direction but also from a crewed platform where normal operations see the manifestation of attitude changes is being investigated for stationary, adrift and underway conditions. The equipment capability, however, has been developed such that additional sensors can be readily incorporated with wave direction being more reliant on streams such as forward and athwartships speed through water alongside apparent wind data.

3.3 CHALLENGES

3.3.1 Quantifying Sea States

A significant challenge is presented as to reliably quantifying sea states at speed in small planing craft using motions data alone. The WaveReader platform has been developed for this challenge and is currently configured to use an internal IMU with both heave and acceleration data.

When stationary the internal IMU allows for necessary data outputs such that data can be quantified in much the same way as a wave buoy however calibration is necessary for both the position within the vessel and the dependant and variable position of the centre of motion of the craft in which it is installed. Whilst some calculation can be carried out with the IMU alone this limits the integrity of the result thus two wired remote accelerometers are made available and used for calibration and to provide a location of the centre of flotation. Furthermore, any centre of motion becomes continually variable once the vessel is underway, significantly if planing when not only does the centre vary significantly through the speed curve but also acceleration and damping effects are to be considered each of which highly dependent on the craft.

There is currently a need for users to input the exact positions of the WaveReader and connected accelerometers as knowing these positions within the vessel is required for a centre of motion be determined. The calculation of this has been seen to be stable in calm water when stationary. The current testing has shown good correlation with wave buoy motions when stationary with output appropriate to the observed and documented sea state, although currently shown during isolated rather than long term trials. Predictions at speed are subject to ongoing development using data from the full breadth of sensor inputs.

Wave period has presented a challenge in the current absence of forward and athwartships speed through water data thus a calibration is undertaken when the vessel is seen to be stationary, either at anchor or when holding station and this is limited to an apparent wave period. It is the intent to refine this calculation with the addition of supplementary sensors however any additional sensors are to augment the onboard calculation rather than present a requirement for it. Such that a philosophy of graceful degradation can be maintained ensuring the equipment remains functional in isolation from any external inputs.

3.3.2 Vessel Interfaces

Challenges experienced in earlier iterations of the analysis equipment showed that whilst systems data for onboard equipment could reliably be collected with the use of specific instrumentation, this presented a considerable installation complexity for trials. As a result, the WaveReader is able to interface with the standardised NMEA2000 data architecture as used by small craft chart plotters, such that any available data feeds can be interpreted including significantly: rudder position and engine speed; and commonly log speed, for which a dual axis input is planned, and environmental factors such as wind data, water depth and water temperature. NMEA equipment is inexpensive commercially available equipment that if not already present on a vessel can be installed and used by the operators outside of the requirement to augment WaveReader data collection.

3.3.3 Simplifying Installation

The current equipment set includes the base station and two remote accelerometers. The remote accelerometers are currently wired to the base station which presents an overhead for installation. In support of the longer-term sea trials of the equipment, NDP are currently developing a full wireless accelerometer device that will be self sufficient for a determined period thus eliminating the need for charge cables for shorter trials.

3.3.4 Data Processing speed

WaveReader has been developed to validate a simulator study in a real-world scenario alongside moving the available research capability, from simple data collection, post-processing and analysis; forward to a system capable of real-time processing of data streams and rapid calculation allowing for customisable predictive outputs to be utilised onboard by crew and coxswains.

The speed of processing has been a concern for the development, as such the time complexity of the software architecture has been minimised and the architecture optimised for the processor cores. This has shown satisfactory performance at sampling rates used for wave motion analysis in the order of 40Hz, the goal is to allow data capture at WBV sampling rates which require some 10-fold the current speed. This use case has yet to be fully tested however current use of lower capability processors allows for appropriate upgrade should the need arise.

3.4 USE CASES

WaveReader has been developed not only as a concept helm aide but also as a versatile and fully customisable motion analysis platform that can now easily be developed to interface with any number of input devices available on sea going craft and provide the capability of real-time data analysis with, currently a customisable GUI offered via a Wi-Fi connected front-end accessible from any suitable device – smartphone, tablet, laptop etc.

The inherent requirement to characterise motions utilises similar measurement devices as employed with scientific motions analysis, thus as a fully proprietary tool that can be used not only across the maritime fleet, but also for land and air domain analyses with profound effect. As an open architecture and a non-vendor-locked tool, WaveReader has been fully customisable to MOD requirements, and not only does it make progress towards the helm aide requirement and future augmentation to the safety case for autonomous vessels, but is also now a refined processing platform that can easily be developed and customised for the benefit of a breadth projects. Current potential use cases that are being investigated include:

- Realtime Calculation tool
    - Helm Aide and feedback
- Motions Analysis Data logger
    - HSC Motions Analysis
    - N&V Data Collection
- Digital incline tool with live outputs
    - Expected reliability improvement for exposed pendulum locations
    - Testing planned
- Environmental Monitoring
    - Doesn’t have to include IMU for motions analysis

COTS boat sensors are robust and reliable
Customisable to client requirement

- Static Wave Buoy
  Deployable in any floating structure
- Future Autonomous applications
  WaveReader also presents a capability that augments the developing technology systems for autonomous operations by providing live indication of sea state from motions analysis allowing for a statistical determination of expected impact severities. As such, given a user input concerning the level of risk that an operator is willing to accept, the vessel operation can be kept below a threshold to protect equipment and / or payload items.

5. CONCLUSIONS

The WaveReader equipment fills a gap in motions analysis that the NDP identified and has significantly improved the equipment set available for its HSC motions analysis research. The focus and intent to provide a helm aide for high-speed operations is under active development for the validation of the simulator trial and future safeguarding of boat crews, passengers and equipment. Isolated trials can be limited in their ability to provide robust conclusions and hence the long-term testing of the WaveReader device will allow the simulator outcomes to be proven in a real-world scenario.

The key conclusion from the preceding simulator trials was quantifying the impact that controllable changes to operations have on the aggregate and instantaneous scale of exposure to boat motions. Exposure to boat motions has been quantified in TTLV and VDV and it has been shown that a speed reduction of some 5% from top speed can equate to a reduction in VDV of some 20%, this relationship can be seen in Figure 5, indicating that planners could significantly reduce WBV risk to crew with a modest reduction in average speed.

What has however been apparent throughout a wide range of trials is that with any mitigations towards exposure to WBV effects is that mitigations should be used to reduce the impact on personnel and equipment and to increase efficiencies rather than providing safeguard for increasing overall operational speed.

6. REFERENCES

1. BRITISH STANDARDS INSTITUTE., 'BS ISO 2631-1, Mechanical Vibration and Shock – Evaluation of Human Exposure to WBV', 1997

2. LLOYD., 'Seakeeping – Ship Behaviour in Rough Weather', 1998

3. LEWANDOWSKI., 'The Dynamics of Marine Craft – Manoeuvring and Seakeeping', 2014

4. STEELE et al., 'Wave Direction Measurements using Pitch-Roll Buoys', 1992

7. AUTHORS BIOGRAPHY

Shaun McCance MRINA holds the position of Senior Consultant at Frazer Nash Consultancy where he is seconded to Naval Design Partnering specialising in onboard technology systems. The Author is responsible for the development of the WaveReader Equipment and its implementation on test platforms. The Author's previous experience includes an extensive power generation and machinery control systems upgrade onboard a major vessel class alongside concept design, engineering analysis and a breadth of experience as a software developer.

Dr Tim Newman held the position of WBV Consultant at NDP undertaking the MARIN simulator trial and the analysis of the results from this.

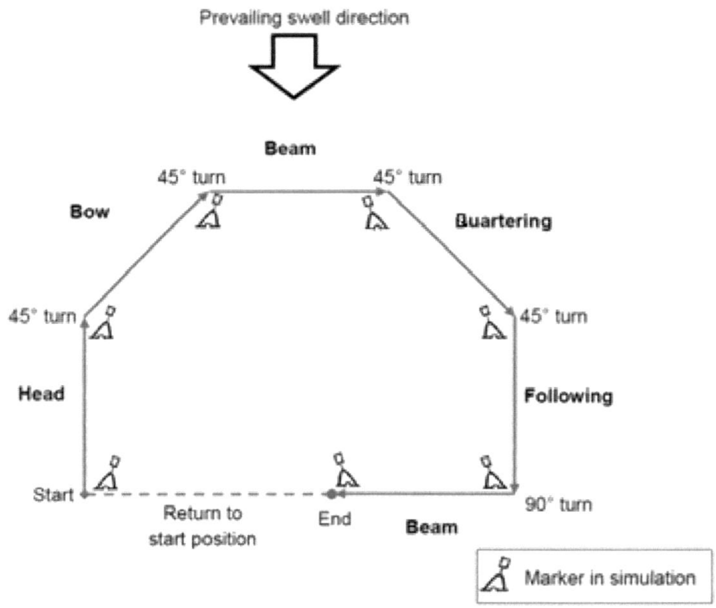

Figure 1: Simulator Sea Trial Pattern

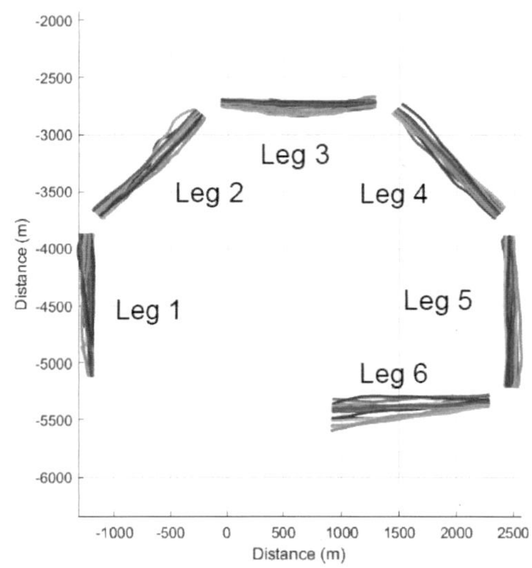

Figure 2: Rationalised Data Sets used for Analysis

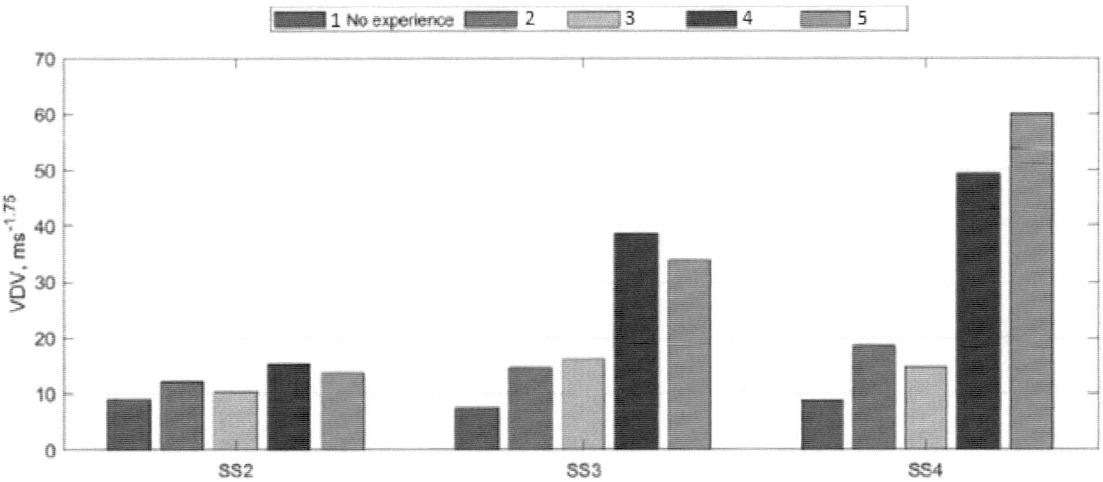

Figure 3: VDV for Coxswains order of experience

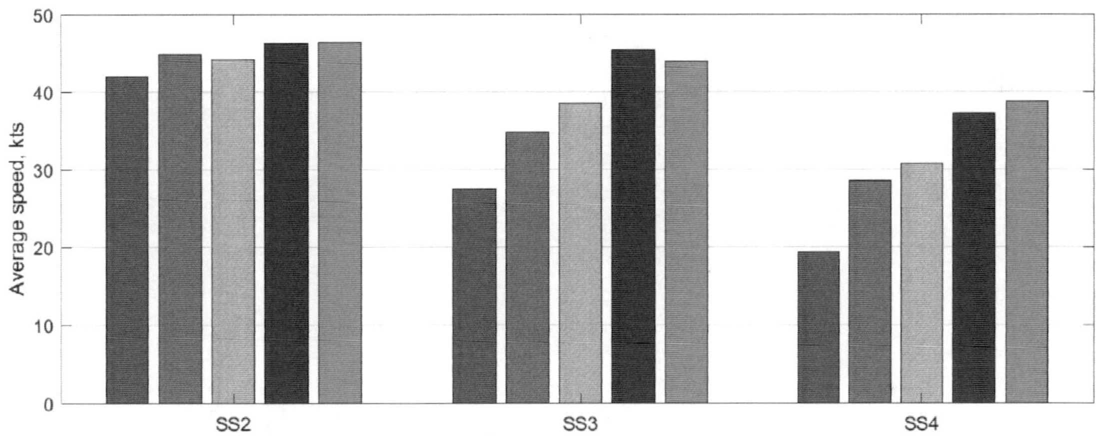

Figure 4: Transit Speeds achieved by Coxswains in order of experience

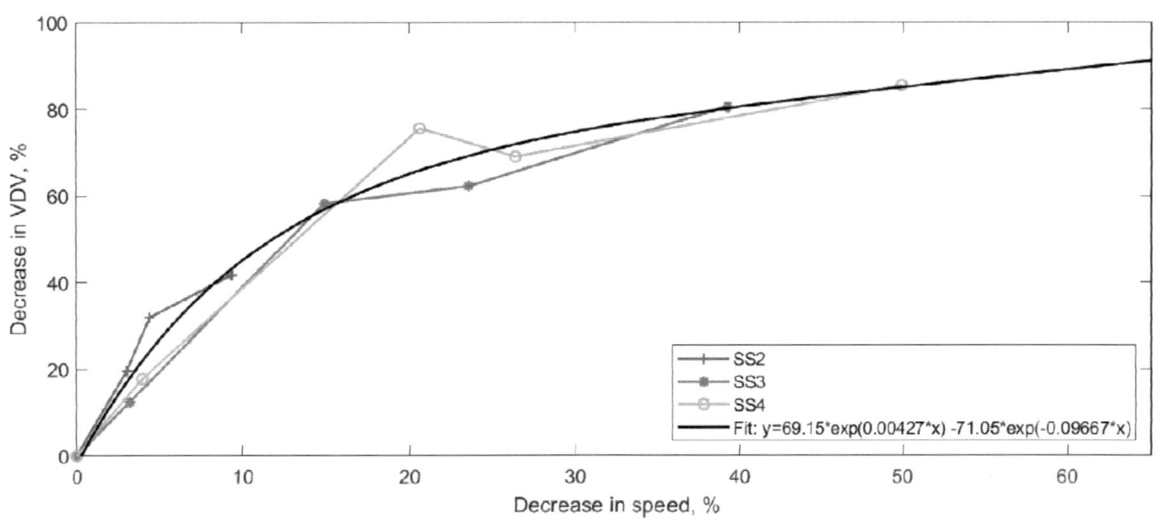

Figure 5: Benefit of Reduced Speeds

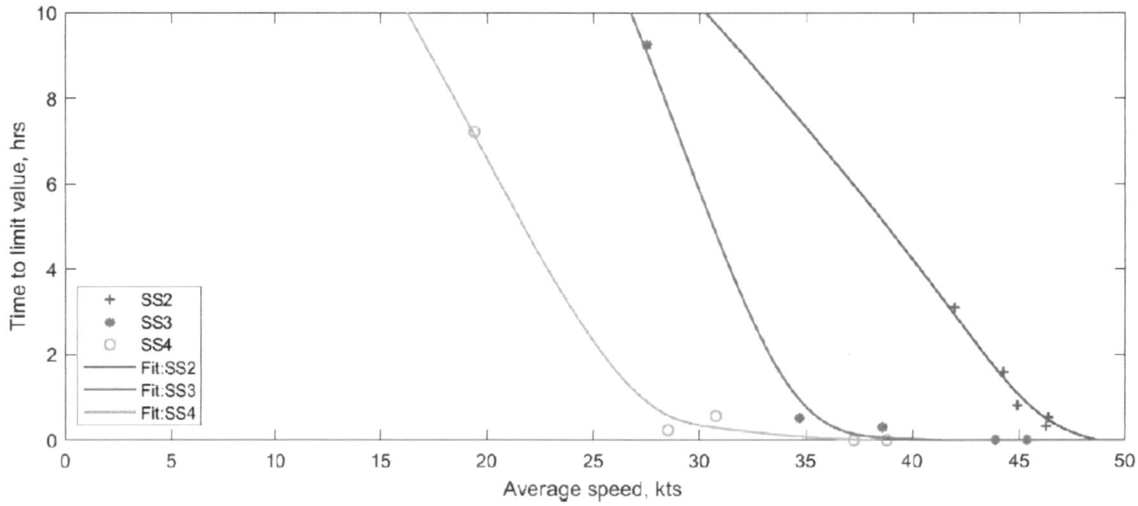

Figure 6: TTLV related to Speed

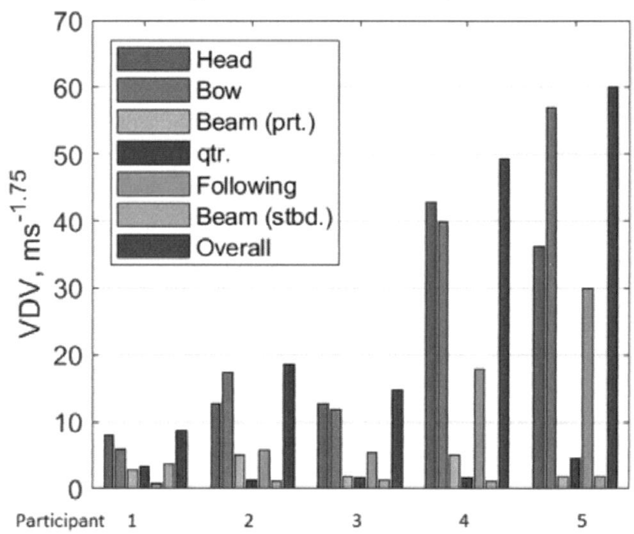

Figure 7: Sea State 4 VDV

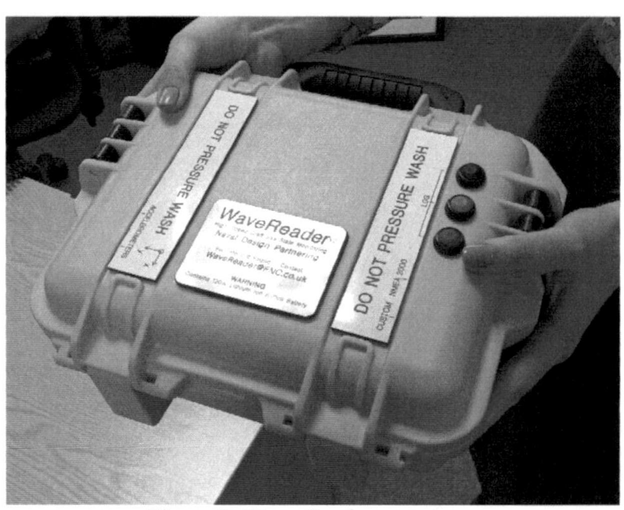

Figure 8: Wave Reader prototype

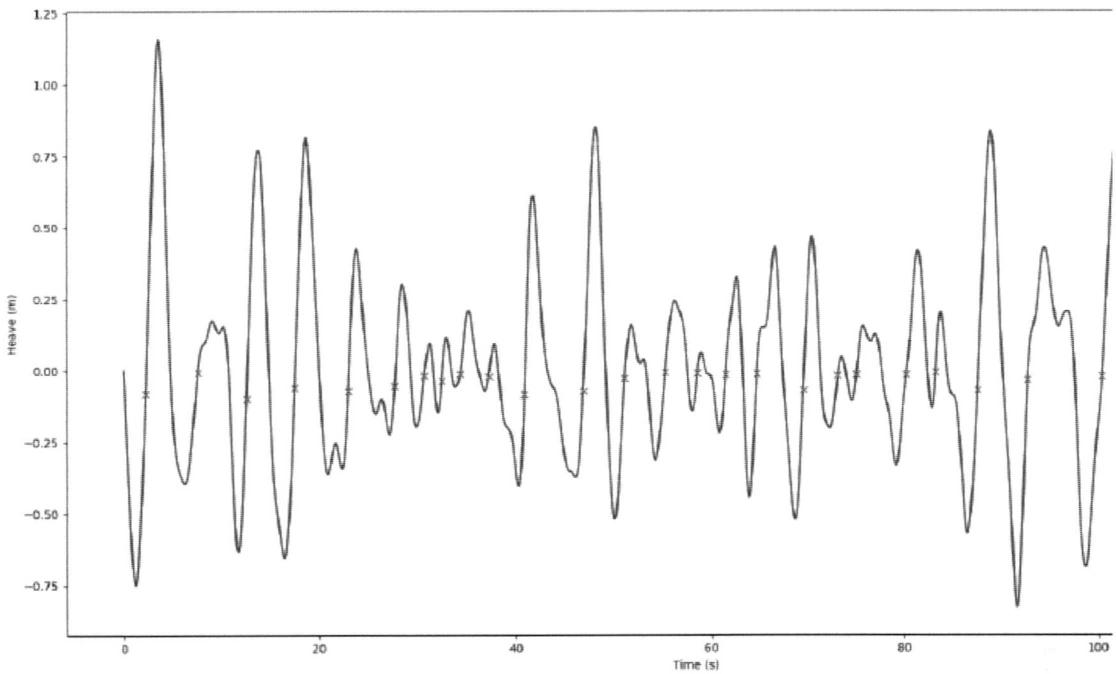

Figure 9: Heave signal as a function of time

A ZERO-CARBON OFFSHORE VESSEL DESIGN

S J Weatherley, R J Sime, A L Page, and **R M Cripps** Chartwell Marine, UK

SUMMARY

Decarbonising the marine industry is challenging but critical.

Although there is significant ongoing investment and research into low- and zero-carbon fuels there is still little consensus as to which, if any, of the current leading contenders will eventually be adopted as the defacto "fuel of the future" for the maritime sector. Clearly the marine industry presents many distinct challenges when compared with fuelling on land, most especially the long-distances involved and lack of charging infrastructure.

In this paper we present a design study for a zero-carbon, long-distance, offshore-capable leisure vessel, fuelled by a single-fuel: methanol. Methanol is one of the future fuel contenders and – certainly for the leisure marine industry – it presents some obvious benefits over alternative fuels.

We explore the vessel design, the propulsion options, safety challenges, and the environmental benefits of such a vessel when compared with traditional diesel fuel.

1. INTRODUCTION

At sea, decarbonisation presents obvious challenges due to the long distances, adverse environmental conditions, and lack of available charging infrastructure. This means vessels must carry enough fuel-energy for the duration of their journey, without the ability to refuel or recharge.

We are in a period where the marine industry needs to change, but is in the process of understanding the benefits and downsides of moving away from traditional fuels. There are a number of competing "clean" marine fuel alternatives at present, with the most widespread being: electrical storage (batteries / solar), hydrogen, ammonia, LNG and methanol.

Regulation and price may eventually determine the direction of the commercial marine sector, as safety considerations and complex storage issues can be built into ship design, safety procedures, and crew training. Within the leisure marine sector it is potentially a simpler landscape with less regulation, smaller boats, and a lack of mandatory training: this means the fuel of the future has to be both easy to use, readily accessible, and safe by default.

Whilst there has been progress in developing pure electric and hybrid diesel-electric vessels, if we are to successfully reduce emissions from the leisure marine industry the answer lies beyond (current) battery technology. The energy-density of leading Li-ion batteries is approximately 30-50x less than diesel, presenting significant weight issues for vessels needing to travel a reasonable distance without recharging infrastructure.

Methanol is a liquid at room temperature and pressure, meaning it can be stored and poured easily. It presents little in the way of environmental hazards as it easily disperses in water, and for humans it can be safely handled with relatively few precautions. Most importantly perhaps, it is also an energy dense molecule, and can be produced from sustainable (carbon neutral) sources.

Methanol as a "feedstock" for storing and producing hydrogen safely (via reforming technology and fuel cells) promises a way of harnessing clean technology that can be used for long-distance operations. Methanol can also be used in (modified) combustion engines potentially providing a cost-effective transition from diesel / gasoline operation to a future fuel-cell-based, cleaner alternative.

In this design study, we explore the design for a small (<24m) leisure vessel, powered by methanol via reforming fuel cell operation, and via injection into an internal combustion engine. This platform will be hydrostatically representative of larger commercial vessels which will likely become technology adopters in the future.

The various technologies explored are all currently available commercially; however they are largely untested in small craft applications. The available equipment provides difficult challenges, including but not limited to: weight, geometry, carbon capture, effective engineering, and efficient conversion to energy.

Whilst the paper outlines the design for an explorer leisure vessel, the overall design and desired use of alternative fuels equally apply to other applications in the leisure, commercial, search and rescue, and military sectors.

2. FUELS OF THE FUTURE

The quest for a low- or zero-carbon fuel is not an easy one. Each potential fuel has its benefits and issues and we will briefly explore each in turn. The major fuels being explored by the maritime sector at this time can be broken into five major contenders: hydrogen, liquified natural gas (LNG), ammonia, methanol, and electrical storage (currently Li-ion).

Clearly one of the most important aspects of any fuel is its energy density (i.e. how much energy can the fuel store for a given weight or volume). The energy density allows us to understand roughly how much fuel we are likely to carry when compared with the defacto standard.

Energy density is most often expressed in two forms: gravimetric and volumetric. Gravimetric refers to the energy density per unit mass of a fuel, whereas volumetric refers to the energy density per unit volume. Clearly both are important when we talk about vessel design. As a good example, hydrogen has a very useful energy density per unit mass (c. 40kWh/kg) when compared with diesel (c. 13kWh/kg); however to store 1kg of hydrogen at standard temperature and pressure requires around $11m^3$ of storage volume. This is where volumetric energy density becomes more relevant, and we should examine necessary gas storage conditions. As an example, hydrogen stored in relatively standard commercial storage conditions (compressed to c. 70MPa) has an energy density of around 1kWh/litre, compared with diesel at around 11kWh/litre.

Below are some approximate energy densities (gravimetric and volumetric) for the most likely fuels of the future.

Table 1. Approximate Gravimetric and Volumetric Energy Densities (combustion) of the major fuel competitors. Values are approximate and taken from multiple sources online.

Fuel	**Gravimetric (kWh/kg)**	**Volumetric (kWh/litre)**	***Comments***
Diesel (reference)	**12.7**	**10.7**	
Gasoline (Petrol)	12.9	9.5	
Hydrogen liquid	39.4	2.8	*Liquid at -253C*
Hydrogen - compressed 70MPa	39.4	1.3	*"standard" storage conditions*
Hydrogen (ambient conditions)	39.4	0.003	
LNG	14.9	6.2	*LNG stored at -160C*
Ammonia	5.2	3.2	*Ammonia stored at -33C, or high pressure*
Methanol	5.5	4.3	*Liquid in ambient conditions*
Li-Ion	0.2-0.4	0.3-0.7	*Approximate current bounding values*

Energy density is obviously not the only relevant topic. Once we have found a fuel with a useful energy density for the passage required, we have to consider safety, storage and handling requirements for that fuel and how those factors impact our vessel design.

Unfortunately, when we combine the energy densities, with the storage requirements, and safety consideration of most of these future fuels it is easy to see why the marine industry has not yet converged on a single obvious solution.

Table 2. Qualitative energy, storage and safety of the major fuel competitors when compared to the diesel reference.

Fuel	Energy density	Ease of Storage	Safety	Comments
Diesel (baseline)	Very Good	Very Good	Good	Easy to store with high energy density, low flammability and toxicity. Vapour presents little risk.
Gasoline (Petrol)	Very Good	Very Good	Moderate	Easy to store with high energy density, moderate flammability (low flash point), and low toxicity.
Hydrogen - compressed 70MPa	Moderate	Poor	Moderate	Hard molecule to store, with moderate energy density. Large, heavy storage infastructure is the dominating issue. High explosion risk under leak conditions. No carbon atom present so excellent for carbon neutrality.
LNG	Good	Moderate	Moderate	Storage under high pressure and/or low temperature necessary, but containment easier than hydrogen. High energy density compared with other fuels, and easy to produce. Poses explosion risk under leak conditions. Most relevant is that LNG cannot be produced carbon-neutral.
Ammonia	Moderate	Moderate	Poor	Storage under high pressure or low temperature necessary. Moderate energy density. Highly toxic if leaks occur. No carbon atom present, so excellent for carbon neutrality.
Methanol	Moderate	Good	Good	Moderate energy density liquid under standard conditions. Low flashpoint and colourless flame provides a safety challenge. Low toxicity compared with ammonia, and low risk to marine life because of dispersion.
Li-ion	Poor	Good	Good	Very poor energy density when compared with other fuels; however under the correct use parameters provides a high-safety, zero-carbon solution. Small danger of "runaway" fires.

Although the table above shows some of the major aspects of the fuels, it is worth bearing in mind that – for the purposes of this paper – we are focusing on a fuel that can be easily used by the leisure marine industry. When considering all of the factors above, it is likely that the leisure industry – if forced by legislation or reputation – has really only two fuels to consider: methanol and Li-ion.

Methanol is a globally traded commodity, available from tankers in most countries around the world. Unfortunately, at the time of writing, the vast majority of methanol is not produced sustainably and is created from either coal or natural gas, meaning its "well-to-tank" emissions are highly relevant. Bio-methanol (produced from decomposing biomass), or e-methanol (produced from carbon capture) are the two major carbon-neutral versions of the fuel. Their production is growing and reducing in price as investment increases alongside demand. Despite the lack of "sustainable" methanol production, there are environmental benefits of using this fuel whilst investment stimulates production of the carbon-neutral version.

3. POWERING OPTIONS AND FUTURE-PROOFING

There are two basic routes in which methanol can be used as a fuel for producing energy. The first is to simply burn it in a (converted) internal combustion engine (ICE); this technology has been been used on land (e.g. Indycar racing) and is tested technology, albeit not yet used in the marine industry. The second is to use the methanol as a "feedstock" for producing hydrogen: the methanol is "cracked" in a reformer to produce hydrogen which is then fed into a fuel cell to produce electricity. There are clearly advantages and disadvantages of each method of convering methanol to useable energy.

Within the marine industry there is currently little investment in producing small (<1MW) methanol engines. There is widespread investment in producing larger pure and dual-fuel methanol engines for commercial ships, but clearly the economies of scale mean that there are only few manufacturers concentrating on small methanol-fuelled internal combustion engines suitable for an application of our size. One such company is Scandinaos AB / Enmar in Sweden who produce two sizes of methanol ICE (13 litre, and 16 litre). The advantages of a methanol ICE include the ability to drive straight shafts through a standard gearbox, reducing potential electrical losses, alongside the trusted structure of a combustion engine. Equally these engines are IMO Tier 3 compliant by default, and produce no SOx and around 30% NOx when compared with a diesel engine.

Reformers on the other hand are a relatively new technology and being improved and developed to meet growing market demand. The advantages of a reformer / fuel-cell combination include quieter, vibration-free operation, and (in the future) the possibility of capturing the CO_2 produced as a byproduct. The disadvantages include the high technical complexity, weight, geometry, and operational sensitivity to angle of heel / vibration; this is before even considering the current cost of such units.

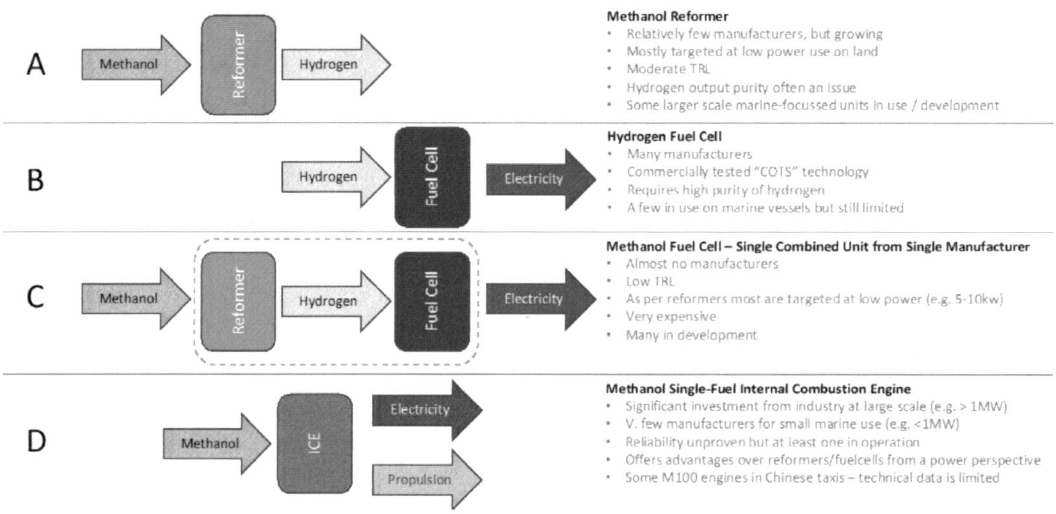

Figure 1: Diagram showing the main routes to energy from methanol, either A+B (from multiple suppliers), C (single supplier) or D

As our vessel is likely to be an early adopter of all of these technologies, and because the technologies are adapting and changing at a rapid pace, we are focussed on ensuring that the vessel is as "future-proof" as possible. This allows the design to be adapted and developed alongside the rapid development, and probable falling cost of the various methanol technologies.

As we can see from the table below, at the time of writing the weights and capital costs of reformer / fuel-cell technologies are prohibitive when compared with methanol-ICE, and certainly compared with diesel.

Table 3. Very approximate comparison of methanol power weights, initial capital costs and volumes compared with a diesel reference

Technology	Approx price per kW (£ GBP/kW)	Approx weight per kW (kg/kW)	Approx volume per kW (litres/kW)	Comment
Diesel (reference)	0.1-0.2	2-3	2	Based on 4 cyl 4.5l compact marine diesel
Methanol-ICE	1-2	5-6	8	Based on converted 6 cyl 13l methanol ICE

Methanol reformer + fuel-cell	5-7	15-18	50	Based on a 300kW reformer + fuel cell combination

To this end our design focuses on a combination of technologies.

- Methanol-ICE for the main (high-power) propulsion
- Parallel hybrid / Li-ion for auxiliary low-power propulsion
- Methanol reformer + hydrogen fuel-cell for the hotel loads, and auxiliary propulsion charging
- Solar cells for extra "free" energy

This combination of energy producing technologies allows us to future-proof the vessel as any one of these techologies improves in either cost, weight or geometry.

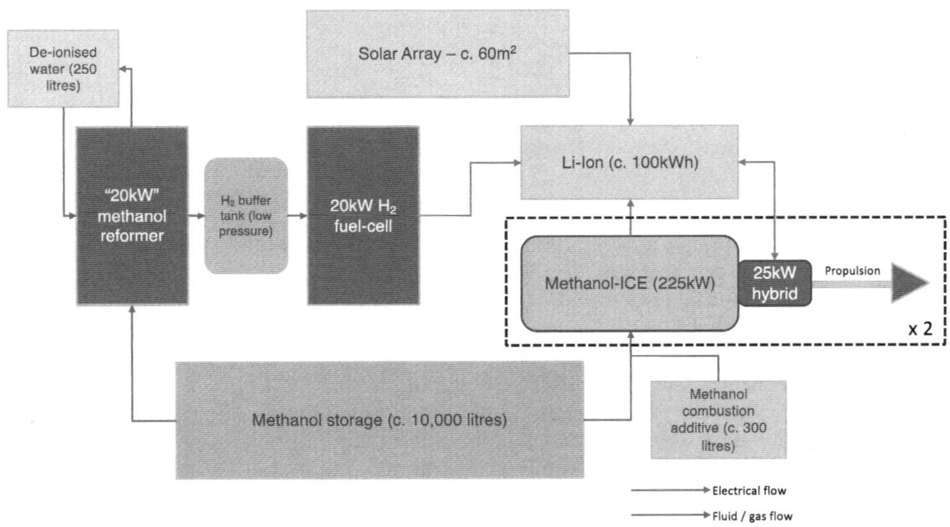

Figure 2: Simplified schematic showing the powering combination for the methanol-fuelled vessel

4. AN ATTRACTIVE, EFFICIENT, SAFE VESSEL

Clearly for any leisure vessel, one of the most important features is the styling of the vessel. From the outset we aimed to design an appealing modern boat which departs from traditional GRP leisure vessels, and instead offers a blend of modern lines, clean aesthetic, and a commercial robustness found in explorer vessels.

The long-range adventure and explorer leisure yacht market is currently growing. One of Chartwell Marine's current leisure vessels (the Chartwell Explorer 47) has seen commercial success sold under the brand name Archipelago 47. Our working assumption is that this explorer market is going to continue to grow, but also likely to see a shift into carbon-neutral, and environmentally conscious fuelling. This will be based on the demands of the owners, but also shifting environmental legislation – especially around some of the more highly-prized adventure desinations (e.g. Norway, Svalbard, South Pacific islands etc). Equally the "adventure" and longer-range leisure market is one where Li-ion and/or solar doesn't offer a viable alternative to diesel, because of the distances travelled and the potential higher latitudes (and hence low solar radiance) involved.

For these reasons, for the zero-carbon platform we have opted for a lengthened version of our original 47-foot explorer vessel, based on the same styling and efficient hull form. Based on a number of economic and design criteria we have settled on a vessel length of 63-foot (19.2m) with a beam of 28-foot (8.5m).

The longer vessel will achieve a number of benefits:

- Appealing to a higher-net-worth audience more focussed on long-range cruising, and hence the move to methanol is less of a % increase in vessel cost

- Better able to accommodate the likely weight and geometry of methanol propulsion
- Increased waterline length and hull separation providing a more efficient hull form.

Figure 3: Initial artists sketch of the Chartwell Explorer 63

4.1 HULL FORM

The Chartwell Explorer 63 platform uses a catamaran hullform derived from Chartwell Marine's proven aluninium catamaran workboat designs. A catamaran form is an interesting alternative to a monohull for a small explorer yacht, as it offers a more stable platform, larger main deck area and a large deck house roof area to carry solar panels. Furthermore, the separate machinery spaces offer greater redundancy in the event of damage to an engine room in remote locations, and isolation of the methanol systems. A catamaran platform also allows the use of two very slender demihulls offering high-efficiency operations and low fuel-burn characteristics in calm water and in waves, and provides a comfortable ride in a variety of sea conditions. These are important considerations for a methanol powered yacht.

A long range capability at low speeds (5-15 knots) is the primary performance target, but the vessel must also have the ability to operate at higher speeds for short periods (20+ knots). Throughout this speed range a catamaran with a high length-displacement ratio (i.e. long and light) will offer lower resistance characteristics, lower powering requirements and lower fuel burn in comparison to a catamaran with a low length-displacement ratio (i.e. shorter and/or heavier).

The Explorer 63 catamaran hull form was initially scaled from the existing Explorer 47 catamaran form which is optimised for speeds greater than 20 knots; however, the Explorer 63 catamaran is primarily intended to offer long range capability at speeds below 20 knots, with high speed performance being a secondary requirement.

Several hull design variants were developed and assessed through a series of Computational Fluid Dynamics (CFD) studies to optimise the hull for low speed, long range operations. These hullform optimisation studies focussed predominantly on changes to the after body design and transom immersion, but also included some minor re-fairing of the forebody.

Further variants were assessed to determine the effects of moving to round bilge displacement form rather than a hard-chine semi-displacement form. The final optimised hull model was designed to accommodate a conventional fixed pitch propeller drive train, or a controllable pitch propeller.

4.2 PERFORMANCE

Based on our weight estimate a powering curve has been derived using CFD analysis. This verifies that our choice of powering for the Explorer 63 is sensible, and suffient. The low-speed curve below, shows that our installed twin 25kW electric motors should allow a peak speed (on departure) of around 9 knots. Equally the long-term, reformer-powered, electric silent running (20kW total) should allow 6-7 knots.

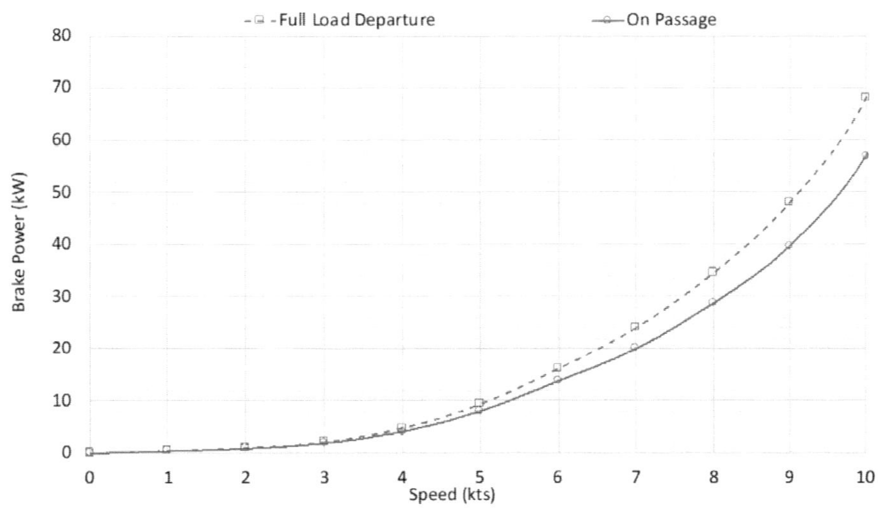

Figure 4: Low-speed powering requirements of the Chartwell Explorer 63

Where we move into the higher-speed regime, and over the hump of the vessel our installed twin 225kW methanol combustion engines will give a top speed of around 19-20 knots, with an efficient cruising speed of 15 knots.

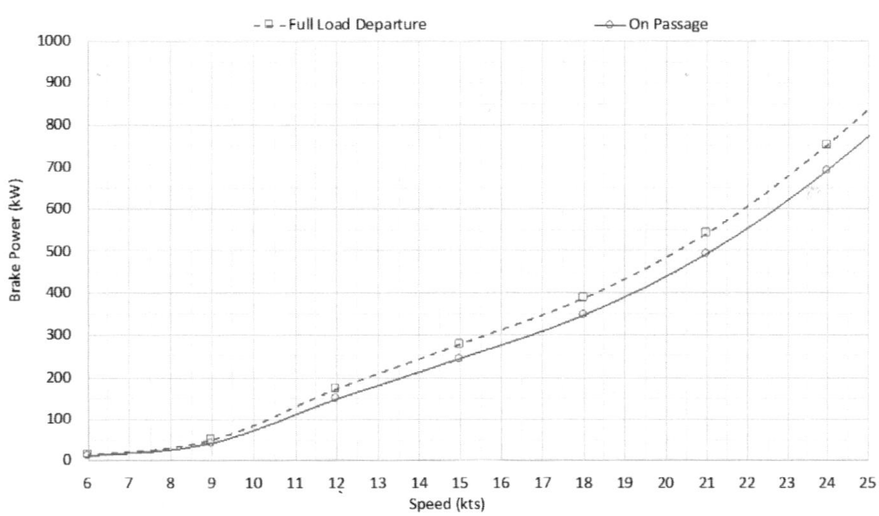

Figure 5: Higher-speed powering requirements of the Chartwell Explorer 63

4.3 RANGE

Clearly range is a key factor when both designing a capable explorer vessel, but equally when investigating methanol propulsion. With methanol's energy density being approximately 50% that of diesel we should expect – as a very simple approximation – half the range of a pure diesel vessel.

Below is the range curve of the methanol Chartwell Explorer 63, based on a deadweight assumption of 10.5 tonnes, including 10,000 litres (7.9 tonnes) of methanol.

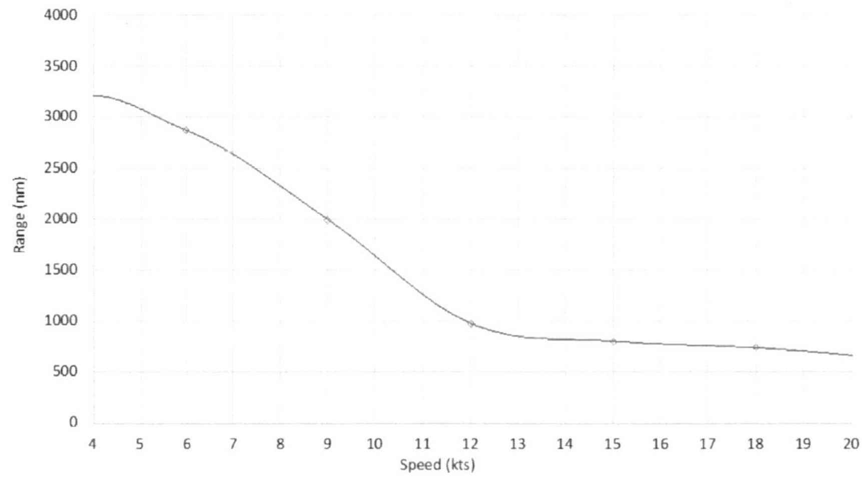

Figure 6: Approximate range calculation of the methanol-fuelled Chartwell Explorer 63, based on 10,000 litres methanol fuel carried, with sea margin, and 10% reserve fuel. This does not take into account range extension from solar / electric propulsion.

Figure 7: Final render model of the Chartwell Explorer 63

5. SAFETY CONSIDERATIONS

As stated in section 2 it is very likely that the leisure marine industry will be subject to less strict operating procedures and licencing than the equivalent commercial vessel. This means it is paramount that the vessel designer and builder takes into account an ALARP approach to risk, reducing the potential dangers of operating a vessel with a low-flashpoint fuel such as methanol. Clearly appropriate documentation of potential hazards, and procedures for routine maintenance would have to be provided in any owners manual for the vessel. Equally a training course for potential owners could be a valuable, and pertinent addition to any vessel sale.

We are currently engaged with Lloyds Register in a HAZID process to ensure that the appropriate safety requirements have been considered. Although this process is not yet fully completed it is relevant to talk about some of the major aspects of methanol fuelling which need to be considered for a vessel such as this.

- General arrangement and layout: Our vessel will only feature one potentially hazardous accessible area within each hull, and this is the engine room. Within the engine room we intend to contain all of the methanol systems including ICE, reformer and fuel cell. This allows us to suitably fire protect, and ventilate a single space within each hull. This keeps construction simple, and safety maximised – ideal for the leisure industry.

- Tankage: Methanol should be stored in stainless steel tanks rather than integral aluminium tanks. It would be preferential to contain the main fuel tanks within a cofferdam, with ventilation and vapour detection to reduce the risk caused by a potential tank leak; however one of the benefits of a catamaran is that we have exposed space under the bridge-deck of the vessel, and we intend to use externally mounted belly tanks, reducing any risk of vapour leaking into accomodation spaces. These tanks will have all filling / venting / sensing equipment accessible through the bridge-deck of the vessel.

- Pipework: All of our pipework will be double-walled pipework leading to engines and methanol control systems. Suitable vapour detection will be fitted to the double-walled pipework to detect and leakage from the inner pipework. Isolation valves should be fitted at regular intervals to minimise any methanol exposure when changing filters / replacing damaged pipework.

- Ventilation and vapour detection: Clearly all methanol spaces should be suitably ventilated using fans and vapour detection equipment suitable for low-flashpoint fuels. Methanol vapour is slightly more dense than air, so consideration should be given to low spaces in potentially hazardous areas. Equally when methanol spaces are ventilated to deck, the pipework should be high enough to allow suitable dispersion before reaching the working deck, and hazardous areas marked out on the deck space. The alternatives to be considered are ventilation to waterline if the design allows.

- Fire detection and suppression: One major consideration with methanol is that it burns with an invisible, smokeless flame. This is particularly relevant when bunkering, as this is the most likely time for accidental ignition. Suitable infrared detection is necessary in bunkering areas and within the engine room. Equally structural fire protection for the engine room / hazardous spaces would be considered essential. Methanol fires can be extinguished by standard CO2 extinguishers. Equally because methanol is soluble in water, dousing with water is an effective way of controlling methanol fires, and so consideration should be given to fire-dousing pumps.

- Maintenance of systems: Clearly maintenance of major systems, including routine maintenance such as filter changing must be considered. Our design, with a single methanol area within each hull lends itself well to maintenance because all hazardous procedures are carried out in a single well ventilated, vapour monitored area. Appropriate drip trays, and collection of any spill during maintenance should be provided.

6. ENVIRONMENTAL IMPACT

As with most of the fuels of the future, methanol emits CO_2 when converted into energy. This is the same when burned in an ICE or cracked in a reformer.

The emissions of CO_2 per kWh of energy produced are roughly equal between either reformer / fuel-cell energy production, and methanol-ICE energy production. These CO_2 emissions are approximately 0.6-0.7kg/kWh. Equally, because the methanol-ICE is a converted diesel engine, we can compare with the approximate CO_2 emissions when the engine is running as a diesel, which is also around 0.6-0.7kg/kWh. This means tank-to-wake emissions are equivalent to diesel for this vessel.

However, the major savings are when well-to-wake emissions are considered. If using bio-methanol for example there are certain instances in which the well-to-tank emissions are actually carbon negative (when using e.g. decomposing biomass). This means that when using methanol derived from these sources at minimum consider them carbon-neutral fuels.

For a vessel such as ours, engaged on long-distance "exploration" voyages, of perhaps around 1000h per year, running at 12-13 knots (c. 200kW), the carbon dioxide savings could be around 120 tonnes of CO_2 emissions per annum when using renewable methanol.

Further to this a methanol ICE produces no SOx, and little to no NOx so is IMO Tier 3 compliant without any exhaust after treatment. When using the methanol reformer/ fuel-cell there is neither NOx or SOx emissions, combined with almost silent and vibration-free running.

7. CONCLUSIONS

Decarbonisation, or at least carbon-neutrality, is not simple.

If we had only recently recognised diesel as a fuel-source, and were not aware of the environmental impacts of its use, we would be praising it for its high energy density, ease of storage, high flashpoint, and relative safety. Somehow we have to find compromises on any or many of these areas if we are to move away from traditional diesel fuelling in the marine industry.

In the leisure marine industry the landscape is perhaps simplified, as it is unlikely that most consumers are going to be happy with carrying either large volumes of highly-toxic ammonia, or to bear the weight and complexity of either hydrogen or LNG storage; this really leaves only two potential contenders for zero or low-carbon fuelling: batteries and methanol.

At the time of writing, for any boat wishing to travel a reasonable distance, or at a speed greater than low-displacement speeds, unfortunately, batteries do not offer a viable alternative.

We have therefore based our vessel design on the concept of methanol fuelling, to allow long-distance and higher speed operation with a low- or zero- carbon footprint. Methanol is a globally traded commodity, and available in most countries without issue. Bio-methanol is increasing in production levels, but is not yet produced at a scale to make it either sensibly-priced, or easy to obtain; however this is changing and there is a likely pull from the larger shipping industry for more sustainable methanol, which will filter to the leisure industry and smaller-scale boating.

For the time being methanol powering technology exists, and the appropriate safety regime can be followed, to design and create a vessel which will – as we transition to sustainable methanol – provide a benchmark for carbon-neutral operation.

8. ACKNOWLEDGEMENTS

The authors would like to thank Jeff Owen from Blue Deep; and Patrik Molander and Bengt Ramne from Scandinaos / Enmar for their helpful input.

The work in this paper was made possible thanks to an Government-funded, Innovate UK Smart Grant, project number: 10034087.

9. AUTHORS BIOGRAPHY

Dr Stephen Weatherley holds the current position of Senior Project Manager and Innovation Specialist at Chartwell Marine. Stephen graduated from Imperial College London with a PhD in Astrophysics, before moving into business management and cyber security. Stephen has a specific focus on innovation in carbon-neutral fuelling in the leisure marine sector for Chartwell Marine.

Rob Sime holds the current position of Principal Naval Architect at Chartwell Marine. He is responsible for overseeing and undertaking a wide range of naval architectural duties from concept design to delivery, including hull lines development, stability calculations, performance predictions, model testing, surveys and sea trials. Rob leads the technical design development of several innovative vessel designs for Chartwell Marine across leisure, workboat, ferry and defence markets.

Andy Page holds the current position of Managing Director, Chartwell Marine. Andy is an award-winning naval architect who has been at the forefront of UK vessel design since 2011. He recently led a panel at COP-

26 as part of Maritime UK's attendance promoting the use of new technologies to decarbonise the maritime industry.

Professor Bob Cripps is a Non-Executive Director for Chartwell Marine, and technical authority for this Innovate UK Project. Previously Bob was Director of Longitude Consulting Engineers, Director of Engineering College for BAE Systems Surface Ships, responsible for developing BAE Systems academic and engineering reputation as being world class. He was Engineering Manager at the RNLI for 18 years responsible for the design and development of all the classes of lifeboat, launching equipment and moorings.

PROCUREMENT PROCESS FOR A HEAVY WEATHER PILOT BOAT

S Lee, and **C Cain,** BMT, UK
G Phillips, Milford Haven Port Authority, UK

SUMMARY

The safe operation of pilot boats, operating in exposed conditions or in particularly heavy weather, requires that the boats and equipment provided is entirely fit for purpose. When procuring a pilot boat for such onerous operations it is critical that a rigorous and well considered approach be applied such that a suitable capability is ultimately provided for use. MHPA and BMT have worked together to procure a new bespoke heavy weather pilot boat and have evolved a process to achieve this. Technical aspects have been reviewed in detail, starting with determination of the assessable key design requirements and operational characteristics. A tender process has been carried out whereby designers and yards have been provided with a detailed basis on which to fairly compete, and whereby MHPA have been given the tools to make appropriate decisions. In doing so, aspects such as cost-benefit analysis of different features, risk management and appraisal of differing technical solutions have been carried out. Furthermore, the involvement of the crews and the culture of the pilot organisation have been significant elements of the process, such that the vital human element of delivering a potentially high-risk service has been fully involved from the start. This paper describes the procurement approach used, development of key vessel requirements, the incorporation of operator experience, a description of some of the debates held during the process and the lessons learned.

NOMENCLATURE

BaFO	Best and Final Offer
COTS	Commercially Off The Shelf
HFI	Human Factors Integration
H_m	Mean (Average) Wave Height
H_s	Significant Wave Height
LSA	Life Saving Appliance
MCR	Mid Channel Rock
MHPA	Milford Haven Port Authority
TSP	Technical Support Partner
VTS	Vessel Traffic Services
WBSV	Whole Body Shock & Vibration

1. INTRODUCTION

The focus of the paper is on the preparation and implementation of a successful procurement exercise for a relatively complex small craft. This paper describes the identification of the primary capability and the key influencers for the procurement, as well as the setting of a vision. The importance of stakeholder engagement is discussed, as are the various stages of the procurement programme. The development of measurable requirements against complex operational functions is outlined alongside the approach to prioritisation and scoring. In looking at the full breadth of the procurement process, aspects such as data gathering, trials validation and training are also addressed. It is hoped that the various observations and lessons learned during the process can be of use to other organisations who may be embarking on a similar exercise.

2. MILFORD HAVEN PORT AUTHORITY

The Port of Milford Haven is a leading UK shipping gateway handling liquid, bulk, break bulk and heavy lift cargoes. It is known as the UK's largest energy port and can deliver 30% of the UK gas demand. MHPA is the statutory harbour authority for the Port of Milford Haven. It is a Trust Port and operates on a commercial basis, charging port fees for providing conservancy and pilotage services to the vessels delivering or collecting products at the principal terminals located on the Milford Haven Waterway.

As a Trust Port, MHPA does not have shareholders, and profit is retained for re-investment into the business in support of its key objectives. MHPA is responsible for the safe and efficient movement of all vessels on the Milford Haven Waterway and provides Pilotage, Conservancy and Vessel Traffic Services (VTS).

The current fleet consists of three 19.5m steel pilot boats built in 2006, 2009 and 2016. The two older craft currently provide the port's heavy weather service up to a maximum of 5m H_s and they have undergone life extension programs to maintain reliability in service. The newer craft has an operational limit of 2.8m H_s and is primarily used during summer and calmer periods. A fleet update program is in place to renew the fleet over the next 5 years for both winter and summer service craft.

Figure 1: MHPA Existing heavy weather pilot boat.

3. SERVICE ASPIRATIONS

MHPA have a requirement to increase their current pilotage operational capability with the introduction of two new heavy weather pilot boats. These new assets are intended to increase operational capability to approximately 95%, with boarding operations being conducted in sea conditions up to 6 – 7.5m H_s without the need to undertake under/over carry operations. Under/over carry operations are the need for a pilot to remain onboard the ship until the next port, or the need to get on the ship at an earlier port if weather forecasts indicate conditions where pilot transfer may not be possible. Once proven in service for an extended period they will replace the two-ageing craft for primary winter and heavy weather operations.

To draw on MHPA's 4 core pillars for operations, Excellence, Collaboration, Sustainability and Safety, experience from previous procurements and lessons learned were a significant focus when establishing the procurement process for the new craft. The uplift in service capability is not just provided by the procurement of new equipment as this is only one small aspect to ensure a successful outcome. This has led MHPA to undertake a demanding procurement program, engaging with all stakeholders of the service.

To understand the operational conditions in which the pilot boats operate and to get an understanding of the environmental limits that would provide a minimum of 95% operational capability, a wave study assessment was undertaken in 2020. MHPA operate and maintain a wave radar unit fitted to Mid Channel Rock (MCR) and five years of historical data were assessed to understand the operational conditions and limits. Based on the following statistical data, the current heavy weather craft provides approximately 310 days operational capability with a limit of 5m H_s and to achieve the goal of 95% operational capability, the new heavy weather pilot boats need to be capable of operating with a limit in the region of 6.5m H_s.

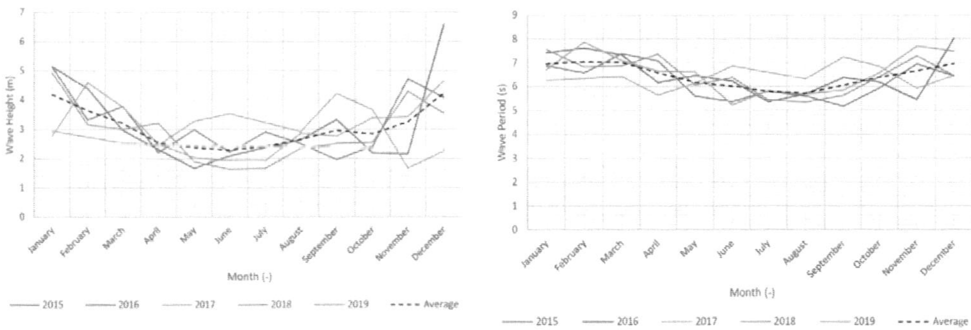

Figure 2: MCR Average Wave Height & Period Data.

It can be noted that the illustrated wave data at MCR is an average wave height and requires a multiplication factor of approximately 1.6 to present as H_s wave height data.

Table 1. Inoperable days vs craft design limit.

Inoperable Days	H_m (m)	H_s (m)
10	4.66	7.46
20	4.07	6.51
30	3.71	5.94
40	3.42	5.47
50	3.19	5.11
60	3.00	4.80

Wave height is not the only metric which influences the operational capability, as wave period can have a significant impact on the wave profile and ability to undertake boarding operations. In practice, short period sea conditions are much harder to operate in than longer periods. The statistical data illustrated in Table 1 should not be taken solely to determine the likelihood of operational success, as operations in greater than 7m H_s may be achievable in long wave periods but may not be achievable in short periods at 5m H_s.

The experience of the pilot boat coxswain is fundamental to determining if a boarding opportunity is both feasible and safe. This takes multiple factors into consideration, including the lee provided by the ship, direction of travel to the waves, weather conditions for visibility, wave period, level of wave confusion and more. Heavy sea conditions present a highly dynamic and potentially dangerous situation where statistical data cannot provide hard limits to operational capability. The coxswain of the pilot boat needs to have full confidence in the operational handling characteristics of the craft to make the operation as safe as possible. There are multiple aspects to the vessel design which attribute to the overall vessel's capability. Even then, it is the Pilot who is undertaking the boarding who will have the final call based on the situation at that time. The pilot needs to have full confidence in the pilot boat and coxswain's abilities, the pilot boat support crew, the characteristics of the boat during the boarding evolution, the access ladder set up on the ship and much more. This procurement exercise focussed on aspects of the vessel design and performance in order to ensure confidence in the boat handing and to the ability to provide safe operations in challenging situations.

In addition to the operational demands of the pilot boat and crew, sustainability is a key factor within MHPAs vision to be carbon neutral by 2030, which needed to be addressed during the procurement process, together with robust training and maintenance regimes to help set new benchmarks for operational excellence.

4. PROCUREMENT ROAD MAP

The procurement process for this project has been extensive and needed to ensure that the correct data is available as the basis for making informed decisions. The following roadmap identifies the general stages of the project and the activities undertaken:

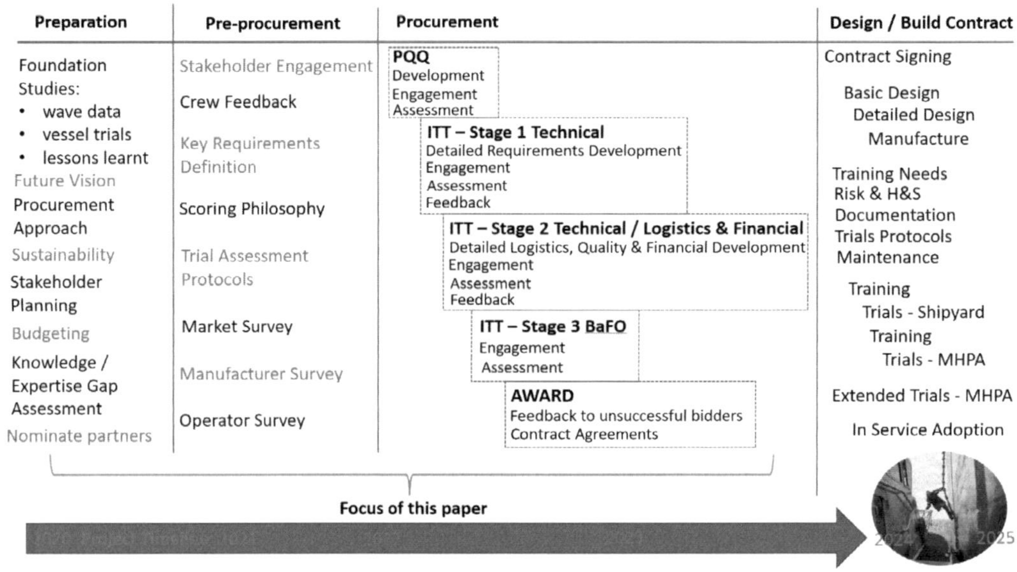

Figure 3: Procurement Road Map / Timeline

4.1 PREPARATION

4.1(a) Foundation Studies

The successful procurement of a new heavy weather pilot boat relates to far more than simply the purchase of a piece of equipment. It is the development of a new capability for the operating organisation, and when considered through this long-term perspective, the importance of establishing a solid foundation for that capability is clear. The procurement preparation stage was carried out over a period of approximately 1 year and included the following activities:

- Development of an understanding of the service requirements
- Review of the support mechanisms necessary to keep the boat in operations, such as maintenance, spares, training and through life support
- Consideration of the necessary extent of redundancy to ensure service reliability
- Programme of stakeholder involvement
- Identification of the environmental conditions for which the service is to be provided
- Multiple sea trials on existing craft to determine the current capability and gaps against expected performance targets.
- Exploration of alternative options including helicopter transfer, moving the location of the pilot boat station or increased under/over carry.

As such not only was a strong basis developed for the procurement, ensuring the right knowledge and data was available to make effective decisions, but also a clear rationale was developed for the service that is to follow.

4.1(b) Stakeholder planning

It was identified early in the project that a very wide range of stakeholders would be involved, and that careful proactive management of differing stakeholders would be critical to the long-term success of the project. To develop an achievable service capability, appropriate input is required throughout every stage of the procurement process. In doing so there has been a consistent shared respect and understanding of the vision of the company from the beginning of the project, as well as of the challenges being addressed by the project team. A stakeholder map was generated as follows:

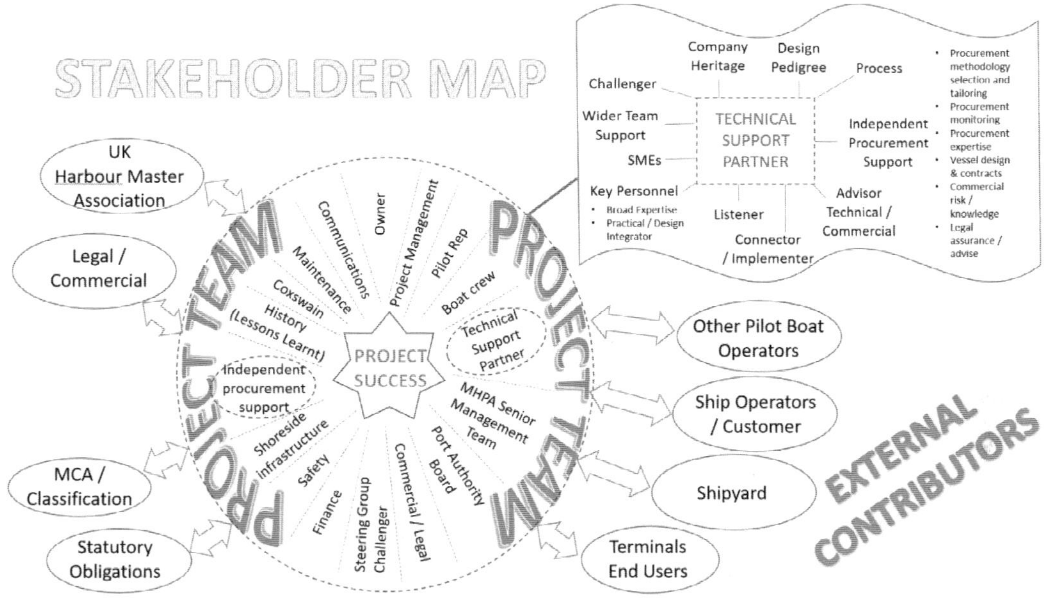

Figure 4: Stakeholder Map

To maximise the success of the procurement, MHPA have focused on utilising the experience and information available throughout the entire pilotage team. The collective knowledge has been used to determine requirements and debate any issues or concerns. The teams operating the craft do so day in / day out and the new vessels will be a part of their daily lives for years to come. Their experience relating to what is required, and what works and what doesn't work, is vital to achieving a successful long-term outcome. All vessel requirements need to be based on first-hand knowledge of the task to be undertaken, and this is particularly the case for the specialist activities associated with pilot boat operations.

When embarking on this procurement programme, MHPA identified that a having a suitable technical procurement support partner could bring significant benefits. In fulfilling this function BMT has supported MHPA with a broad range of activities as shown in figure 04. The attributes of a technical partner are critical to their success in this role, not only in terms of knowledge and experience, but also in terms of behaviours and approach as part of the MHPA team. BMT and MHPA also recognised the value of independence and how this role could support the alignment of a range of potential perspectives from within the MHPA project team. This independence allowed BMT to constructively challenge MHPA at times, whilst also working to support the MHPA management team in their strong aspiration to involve all parts of their internal team. MHPA were keen to make sure that operational aspects and management aspects were approached together. In doing so BMT was able to support MHPA with numerous technical stakeholder discussions, distillation of complex technical aspects from practical experience and analytical data, robust debate on the correct level of design capability, independent perspectives on lessons from previous procurements, and the development of distinct measurable requirements around practical operational activities.

4.1(c) Approach to Procurement

Before embarking on the procurement exercise, the procurement team addressed the approach that was to be taken, looking at the following:

- Review of lessons learned from previous MHPA procurement exercises and from the procurement exercises of similar organisations.
- Consideration of potential investment in the procurement exercise, coupled with a perspective of minimising the long term risks associated with different approaches.
- Time available to carry out the procurement, from project start to the necessary commencement of service.
- Life extension service on existing in service craft

The following list of activities was generated to facilitate planning for the programme overall:

- Crew / Operator workshops / feedback
- Key operational Requirements
- Crew Feedback / Engagement
- Scoring Philosophy
- Technical vs Financial Ratio
- Requirements Definition & Prioritisation
- Risk Assessments
- Trials Assessment
- Timescales
- Market Survey
- Manufacture Survey
- Operator Review
- Shipyard engagement
- Pre-Qualification Questionnaire (PQQ)
- Invitation To Tender (ITT) Preparation
- ITT Competition
- Analysis of Submissions
- Feed Back Process
- Example vessel demonstrations
- Shipyard Assessment
- Down Select Process
- Contracts
- Award
- Design Period
- Build Period
- Trials and Acceptance Period

It is observed that a procurement exercise of this nature involves a significant level of work. The order of activity is also important, as is the need for feedback loops, and the potential for third party related schedule implications. Above all, MHPA recognised that the management, organisation and leadership of the procurement programme was critical to its success.

4.1(d) Budgeting

The budget for the programme is a cornerstone of the preparation for the procurement, and the following aspects were considered in detail:

- Internal budget availability, both as a threshold and as a target
- Realistic consideration of vessel prices, and margins for a range of alternatives
- Consideration of a wider perspective on budget setting, aiming to prepare to strike the right balance between cost reduction, best value and potential implications on vessel capability or adoption of the vessel. At the planning stage it was agreed that the principal focus of the requirement was to be maintained, and that the impacts of any financial savings would need to be understood as part of any decision making. It was recognised as important that all MHPA stakeholders also understood the relationship between restricted budgets and overall capability, targeting smart spending to maximise the final capability.

4.1(d) Sustainability considerations

In the procurement of any new asset, it is a priority not only to make services more efficient in areas like fuel consumption and emissions reduction, but also to act more generally in support of decarbonisation. In approaching this procurement, a robust debate was held to ensure that the balance between this and the principal operational capabilities of the craft was correct. It needed to be acknowledged that more efficient lower power and lower weight vessels were in potential conflict with the overall operational capability, as the boats may not be quick enough to get out of dangerous situations, and the waves could have a greater impact on the vessel motions, making it harder to provide a stable platform to undertake boarding in rough conditions. Through this debate it became clear that the priority for MHPA is to be able to safely put a pilot onto a ship in the anticipated weather conditions, and any performance target that negatively impacts this needed to be carefully addressed. MHPA have a target to achieve net zero carbon emissions by 2030 and aspects of the requirements need to take this into consideration without detrimentally effecting the primary focus of transferring a pilot safely.

4.2 PRE-PROCUREMENT

4.2(a) Key Operational Requirements

To determine a rough idea of the required vessel capability and to assist in undertaking an initial market survey, a set of key operational requirements were developed. This was done in a workshop format with representatives of the operational crew and management team. Key operational requirements are considered essential to providing the service expectations of MHPA's customers, having a direct impact on the principal focus to transfer a pilot. Key requirements were broken down into the following areas:

- Vessel motions, sea keeping and stability
- Propulsion for speed and other operational requirements
- Human factors
- Environmental conditions and influencers
- Personnel requirements
- Through life factors such as maintenance and duty profile
- Regulatory framework

4.2(b) Market Survey

The Key Requirements list was used to undertake an initial COTS market survey, reviewing what was readily available on the market and the capability it could provide, together with a gap assessment against the required level of capability. An initial review of over 100 potential designs was carried out based on aspects such as published maximum operational sea conditions, current in-service applications, speed, structural material, self-righting capability and more. Consolidation of the list identified 10 manufacturers for onward engagement, and a level of informal dialogue provided further insight to the products. This feedback confirmed that there were sufficient manufacturers interested in providing a suitable solution to run a competitive procurement.

4.2(c) Operator Survey

An exercise was carried out to better understand similar existing operators and the vessels that use, such that lessons learned could be applied to this procurement. Discussions with numerous operators highlighted two fundamental risks, firstly the risk of optimising a product for a singular focused aspect, and secondly the risk of losing sight of the principal operational focus of pilot transfer. The relationships built during this phase are also expected to help with the overall adoption of the solution following build and trials.

4.2(d) Definition and prioritisation of technical requirements

An orbit of priority was created to illustrate high, medium, and low priority requirements against which the vessel's equipment and performance requirements would be attributed, assessed and scored. High priority requirements were considered to have a direct link to a primary operational focus; Medium priority were important factors which will have an impact on vessel resilience through-life; Low priority were still important factors to the resilience of the overall operational capability. 'Nice to have' factors have generally not been included in the technical requirements. It can be seen from the orbit that the priority remains the safe transfer of the pilot, and it is around this that the vessel requirements are based.

The development of the list of requirements and their prioritisation was carried out over a series of workshops, and the expertise of the operating crew, maintenance crew, pilots and management team were leveraged to determine what is necessary to provide the required level of capability. This definition is integral to ensuring that the new vessel meets all the areas of operational need. This experience was broken down to determine what is required, why it is required, how it is carried out and how it influences the capability overall. The following were observed during this process:

- The generation of the list itself was relatively straightforward but did require input from the full stakeholder group.

- Significant debate was held on the relative importance of each aspect of the design. This process was iterative, and in some cases difficult decisions were taken to raise or lower the priority of a particular aspect.
- It was noted that due to differing perspectives of stakeholders, importance of requirements inevitably reflected their specific involvement. In carrying out a wide debate of the full range of requirements, a good mutual understanding was achieved across the team.

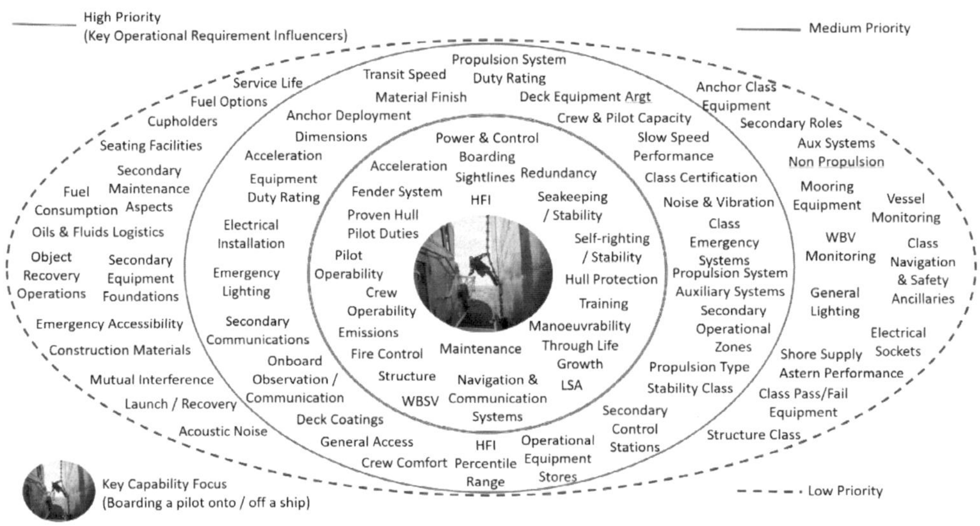

Figure 5: Key Requirements Orbit of Priority

Although many requirements have been listed and debated, the following are considered to be among the highest priorities:

- Pilot Operability – Ensuring safe working area for the pilots to undertake their role effectively whilst onboard the pilot boat and during the transfer process.
- Boarding Speed – Ensuring the pilot boat can be controlled efficiently during the boarding operation, providing a stable platform whilst maintaining a fixed position alongside with opposing external environmental factors.
- Acceleration – Links with boarding speed requirement and is an important influencing factor to maintain a steady position.
- Manoeuvrability – Vessel to be reliably controllable in emergency situations.
- Sea Keeping & Stability – Ensuring the vessel and onboard systems can be effectively utilised in high sea state conditions.
- Crew Operability – Ensuring the vessel can be controlled effectively in all weather conditions and maximising the effective use of the three primary crew onboard.
- Structure – Sufficiently robust to provide high levels of resilience whilst operating in high sea state conditions. Sea wave pressures, hard contact with ship's side and self-righting loads.
- Self-Righting & Stability – Vessel needs to have a level of self-righting capability to keep all persons on board safe during high sea state operations.
- Human Factors Integration – To provide a safe effective operating space for all personnel, vessel motions, control station ergonomics, sightlines, seating, safety rails, hand holds, effective positioning of equipment so not to risk the safety of personnel.
- Life Saving Appliances – To minimise the risk to overboard personnel during recovery operations.
- Through life growth allowance – Ensuring the vessel performance can still be achieved with a suitable allowance for weight increase for through life growth.
- Emissions – Whilst not a strict legislative requirement, emissions are an important factor to MHPA, with the request for IMO Tier III compliant engines and the ability to utilise HVO.
- Maintenance – Engine / major equipment removal, to minimise down time and suitable space / access provide for ease of maintenance and emergency operation.
- Fender System – Multiple characteristics to provide an effective system for the tough arduous conditions which the vessel will operate and ease of maintenance / replacement.

4.2(e) Technical requirements specification

The prioritised list of requirements was then converted into a statement of technical requirements. To maximise the expertise of industry to provide a suitable design solution this route was considered preferable when compared to the alternative of developing a technical vessel solution specification. The technical requirements specification was broken down into the key areas shown in table 2. Over several workshops, functional and measurable requirements were generated under each category. Each of the detailed requirements were assessed against the primary capability focus and priority.

Table 2. Technical requirements category breakdown.

1) Design & Layout	**6)Auxiliary Systems**
1.1 Layout	6.1 Propulsion Engine Raw Water Cooling
1.2 Performance and Operational Profile	6.2 Fuel System
1.3 Logistics	6.3 Bilge System
1.4 Classification / Standards	6.4 Machinery Space Ventilation
2) Structure	6.5 Ride Control Systems
2.1 Design Parameters	6.6 Exhaust System
2.2 Foundations	6.7 Fire Extinguishing Systems
3) Propulsion & Manoeuvring	6.8 Fresh Water System
3.1 Engines	6.9 Black / Grey Water System
3.2 Gearboxes	6.10 HVAC - Air Con / Screen Demist / Heating / Ventilation
3.3 Propulsion / Manoeuvring System	6.11 Lubricating Oil
3.4 Shafting	6.12 General
3.5 General	**7) Fitout**
4) Electrical Systems	7.1 Exterior Fitout General
4.1 General	7.2 Fender System
4.2 24 / 12 Volt DC Electrical System	7.3 Overboard Recovery System
4.3 240V AC Electrical System	7.4 Handrails / Safety Rails
4.4 Lighting	7.5 Davit System- Object Recovery
4.5 Socket System	7.6 Mooring / Towing Bollards & Fairleads
4.6 Hull Protection	7.7 Anchor Arrangements
5)Navigation & Communication	7.8 Windows
5.0 General	7.9 On Deck Storage
5.1 Integrated control and monitoring system	7.10 Wheelhouse Internal Fitout
5.2 Chart Plotter	7.11 Control Stations
5.3 Radar	7.12 Toilet
5.4 External Communication	7.13 Lower Accommodation Area
5.5 Onboard Communication	
5.6 AIS	
5.7 Navigation Safety	
5.8 Vessel Monitoring	

A technical requirements table was generated to form the basis of the technical tender document, with each requirement having the following fields. In generating the table, the intent was to provide a solid and defensible basis for the tenderer to understand each requirement and to demonstrate compliance in a structured manner against which it can be measured by the assessor.

- Statement of Technical Requirements (SoTR) ID number.
- Principal requirement Area
- Brief overview of the requirement
- Threshold and objective targets for meeting the requirement
- Justification for the requirement
- Confirmation of how the requirement will be assessed, such as by analysis or by trial. Where trials are the assessment, a trial protocol needs to be identified.
- Evaluation type
- Priority weighting
- Supplementary information to provide additional background information
- Confirmation of compliance by tenderer

- Statement of compliance for the tenderer to identify the evidence to achieve the claim and to identify the risk level associated with the claim.
- Identify minimum and maximum achievable scores
- Assessor score based on the evidence provided.
- Assessor notes and comments, providing the assessor the opportunity to give feedback against the given score.

4.2(f) Technical scoring philosophy

It was recognised that the scoring philosophy would be fundamental to ensuring a successful outcome to the procurement. Lessons learned have shown how scoring processes can sometimes be very complex and if not handled correctly can significantly inhibit the whole procurement process. Complex requirements and scoring can even deter potential bidders from taking part. However, the heavy weather pilot boat procurement is for an onerous and niche capability and needs a suitable scoring process to ensure the correct calibre of solution. The scoring philosophy was set up as follows, with the aim of ensuring a good balance between promoting the high priority aspects, whilst also recognising that the lower priority aspects were still considered important to the global operation of the craft.

- Requirements were split into Threshold and Objective performance capabilities, where Threshold is assumed to be the minimum level of performance that needs to be achieved and Objective is the ultimate target that would bring further improvements to the overall service provided.
- For Threshold performance, a score of 1 point was awarded. For Objective performance, 2 points were awarded for Low and Medium priority requirements and 3 points for High priority requirements.
- Where a performance can be considered continuous from Threshold to Objective, the points scored would be equivalent to the percentage increase between Threshold and Objective.
- The level of requirement priority Low – High have been applied as a multiplication factor. Low = 1, Medium = 2 and High = 3. The factored score (Objective / High Priority = 3 x 3) would be given 9 points in total.
- For High priority requirements, Threshold capability is considered the lowest allowable performance with rejection applied against anything lower. For Low and Medium level priorities there may be the ability to erode the Threshold level on a case-by-case basis depending on the performance alternative proposed.
- There are also requirements which are deemed Pass/Fail criteria. Without these, the craft would not be certified.

4.2(g) Technical and financial scoring philosophy

MHPA requires a capability to operate at the extremes of reasonable environmental conditions, and this onerous operational requirement forms the backdrop to considering the relative importance of technical and financial scores. The success of increased operational capability brings with it a high financial reward, easing the typical burden of the ultimate CAPEX cost. A percentage split of 70% technical / quality and 30% financial was considered appropriate. This split ratio is aimed to focus the tender solutions to be highly capable, and in doing so was important to have reliable reference data to ensure that value is provided for performance capability.

4.2(h) Trials Assessment

When setting performance targets, it was considered important to provide an outline description of the test method against which it will be validated, and that where a requirement is equipment focused, there is a measure against which to check the final product. A trials acceptance protocol was developed for all at sea performance requirements, detailing the level of information required at each stage of the procurement process. Trials assessment protocols identify the associated requirement which the trial is being validated against, with clear assessable criteria, which will minimise the risk when the craft is built, and trials are conducted. The supply of evidence and the de-risking process begins before the award of contract, with the supply of performance evidence at the ITT stage. Furthermore, during the design stage all associated aspects of a performance measure will be determined to ensure that influencing factors can be managed and assessed through build and into trials.

4.3 PROCUREMENT

Having carried out all the necessary preparation work, the formal tender process was carried out, with the following steps.

4.3(a) Pre-Qualification Questionnaire (PQQ) and initial down-selection

A PQQ process was carried out to allow a formal open process for engagement with industry and to carry out an initial assessment of the interested parties that might be taken forward into the further stages of the procurement.

Together with the key vessel requirements and general background information, a shipyard questionnaire was sent out for completion. This questionnaire provided the first assessment of potential yards and covered a broad band of topics relating to company suitability, all with the aim of reducing risk to MHPA during further stages. Pass/Fail targets provided the framework for a down selection that identified several suitable shipyards to take forward into the next stage of assessment. This process also allowed yards to confirm or otherwise the suitability of their products for the heavy weather conditions stated within the key requirements. The following aspects were reviewed:

Corruption / Bankruptcy
Financial Standing
Capacity & Capability
Quality Management
Equality
Health & Safety
Environmental Practices
Ethical Employment
Retention Bond
Conflict of Interest
Staff Qualifications
Experience
Delivery Capability
Resource Availability
Consortiums.

4.3(b) Invitation To Tender (ITT)

MHPA chose a closed procurement route based on the PQQ responses and sent out ITT information to the selected yards. The ITT stage is typically undertaken over 2 – 4 stages, where at each stage non-compliant or sub-expectation responses can be omitted from further participation and feedback sessions can be held to provide constructive feedback to improve the technical response in the next stage. During the ITT process it was recognised as a positive aspect that tenderers could ask clarification questions to aid in improving their understanding of the requirements. However, this needed to be undertaken in a fair manner, with all tenderers receiving the same information. This process can be very intensive for the tenderers so the level of information requested needs to be commensurate with the reward that they may get in return if successful.

The first stage release of information focused on technical capability aspects of the project, excluding any financial and logistics requirements. It included the Technical Requirements Specification, the scoring matrix and the trials acceptance protocols. The responses and supporting evidence were assessed independently by 3 assessment teams within the Project Team, followed by a workshop to review the results collectively, discussing variations in scoring to determine a single response. Feedback meetings with each tenderer were held to discuss the scoring, allowing contentious points to be elaborated upon and further context provided. At this stage tenderers could be omitted from the further stages, but at this point it was important for a competitive tender that there was a minimum of two bidders, ideally 3.

An important note to consider here for both the tenderer and assessor is that if the compliance evidence is not available, higher mark allocations cannot be awarded. A simple statement that Objective performance will be provided cannot be considered as justification. Simply copying stated equipment and quantity requirements identifies that these will be provided but needs to be considered as a minimum input. Evidence as to what the performance capability of the equipment is and examples of historical evidence will provide increased confidence to award higher marks. For a new project the inclusion of a 'Confidence Factor' would be included in the assessor's scoring, allowing awarded marks to be adjusted based on the evidence provided for all requirements, providing finer refinement on the output score.

Stage 2 also incorporated logistical, quality / delivery and financial aspects. The logistics aspects covered areas such as Documentation, Maintenance, Spares, Reliability, Availability, Supportability, Safety and

Training, with the quality section covering Delivery, Product Pedigree, Yard Capacity, and a physical shipyard assessment.

Technical solutions were updated following the feedback discussions and further information was provided for the Logistics and Quality sections. The physical shipyard assessment was conducted at a slightly later stage during this procurement process but could easily be undertaken at stage 2 to include into the overall stage score. The shipyard assessment provides first hand observations of the working practices, facilities, personnel, and health / safety practices against the information provided at the PQQ stage.

For the financial section each tenderer was requested to break down their overall cost against the secondary section titles of the technical and logistical requirements. This provided the ability to make more granulated comparisons against each tender proposal and allowed for identification of where costs may be high or low. This process worked well in identifying concern areas within some submissions that insufficient allowances had been made to meet the expected level of deliverable. Constructive challenges were presented between the tenderers and MHPA.

The financial proposals were comparatively ranked with the lowest price taking 100% and other prices a sliding scale based on the percentage increase. Following receipt of Stage 2 responses, the assessment process of the technical, logistics and quality/ delivery were undertaken independently, followed by a discussion workshop as per stage 1. Technical scores were updated, and logistic / quality scores were combined. Stage 2 feedback sessions included both technical and financial aspects together with initial commercial, logistical and contract conditions. At this stage, whilst MHPA were aware of the combined technical and financial scores, and the overall ranking, care was required to ensure a fair and competitive conclusion. In doing so MHPA provided some limited information on technical and financial score rankings that could allow the tenderers to take judgements on how their tender may evolve throughout the different stages of bidding.

During stage 2 MHPA took the opportunity to visit operators of boats which the tenderers were putting forward. This allowed assessment of a boat that the proposals are referenced from, to review build quality, and have a vessel demonstration to assess performance. The demonstrator boats were not exactly what was being proposed, but it provided a reference for judgment. The vessel demonstration also provided the opportunity to discuss with the operators, how the vessel handles, what they liked, how the associated tenderer was to work with, things they would change in the future if the opportunity arose. These responses and observations were pivotal to the assessments, providing future links with co-operators for knowledge transfer. Operator to operator discussions can be a powerful tool when assessing complex equipment for complex operations.

Following the completion of stage 2, the final stage 'Best and Final Offers'(BaFO) can proceed, unless there are any changes in the project conditions which would require a further loop to cover any variations. The BaFO stage is solely a financial process where each tenderer is requested to provide their best and final price based on the Technical and Financial rankings provided. There is to be no change to the technical aspects, but the price needs to consider any feedback provided from stage 2.

Once all Stage 3+, BaFO submissions have been received, a combined assessment is carried out to determine the overall rankings, with the contract award process following.

4.4 CONTRACT

This paper is focussed on the preparation and implementation of the tender process. However, in doing so it is acknowledged that the success of the remainder of the project, in particular the design, build and acceptance is directly connected to how the procurement has been set up. The following aspects have been of particular interest.

4.4(a) Design Phase

Once the contract to design and build contract is awarded it is critical to keep the wider operational team engaged during the design phase to continue to transfer their operational experience into the design and arrangement of the craft. Sustained engagement helps to share ownership of the project and will lead towards a successful adoption of the vessel into operational service. The vessel design is expected to be undertaken in full prior to build proceeding to enable a clear agreed path forward with minimal changes during manufacture.

4.4(b) Build / Manufacture

It is important to have the correct level and type of support present through the build phase, such as the following:

- Through build surveys to maintain quality
- Development of detailed trial assessment protocols
- Determine training requirements to expediate full in-service capability
- Documentation to undertake and develop maintenance programs that maximise in service reliability and operational capability.

4.4(c) Training

Training of the vessel operators and maintainers is as important as the craft itself. If the vessel cannot be operated or maintained to its full potential, then there will be a reduction in operational capability.
Training needs analysis is required to determine that the training will meet the needs of the personnel and to ensure that it is applicable to the expected environmental conditions, in particular to maximise capability in rough conditions.

4.4(d) Trials

Trials are the concluding aspect to confirm that the supplied product meets the technical requirements which were set before ITT stage. Some of the tests can be complex and require a consistent repeatable approach to ensure accuracy. The trials protocol will aid in ensuring each trial is conducted in the correct manner and conditions, providing an independent recording of the trials undertaken.

Trials often occur over a specific constrained time period when weather may not be applicable to the expected operational conditions. Therefore, it is important to include an extended period to enable the operators to become fluent with the craft systems and progressively increase the operational boundary in a safe manner.

5. CONCLUSIONS

The procurement of a highly complex heavy weather pilot boat focused on maximising safety during challenging operational conditions requires full team engagement from the outset with a high level of experience across all areas for operational requirements to be captured and articulated correctly as measurable technical requirements. This process should not be underestimated and there will be areas that can be translated in different ways by the stakeholders who are developing a solution to meet the stated requirements.

The applied team efforts in the development of requirements, assessing tender responses, visiting shipyards and operators will hopefully lead to a long, safe, and successful capability for the future.

5.1 LESSONS LEARNT

- Include a ‘Confidence Factor’ in the assessor’s scoring section, allowing awarded marks to be adjusted based on the evidence provided for all requirements, providing finer refinement on the output score.
- Measurable requirements for complex operational needs are difficult to develop and there will be alternative ways to articulate the requirement in a slightly different manner
- Be aware of the workload being imposed on potential bidders - measure the level of information requested against the level of reward available.
- Learn from those who do

5.2 KEY FACTORS

- Have a good spread of contributors to cover the whole operation.
- Ensure requirements are measurable. Some complex requirements may not be measurable, but other less complex measurable performance requirements combined to de-risk / break down the complexities into measurable blocks.
- Keep requirements as simple as possible.
- Identify the key factors which have greatest influence on the primary focus / capability looking to be provided.

- Ensure team engagement is maintained throughout the whole project cycle to maximise operational expertise through to delivery and acceptance.

Hopefully this paper will provide some guidance as to the stages to be considered when procuring complex equipment required to operate in extreme environmental conditions.

6. ACKNOWLEDGEMENTS

The authors thank all operational pilot boat personnel at MHPA for their support during the technical requirements specification development, providing understanding as to how the vessels are used and challenges faced during daily operation. Without this fundamental information, measurable requirements development cannot occur. The authors would also like to thank MHPA management for providing the opportunity to discuss their experiences to date during the development of this procurement process for their heavy weather pilot boat and sharing their operational vision.

7. AUTHORS BIOGRAPHY

Steven Lee is Lead Naval Architect for the Government & Security sector within BMT Specialised Ship Design. His broad base of experience throughout all principal design areas has been developed over the past 30 years during the design of; Military High Speed Landing Craft, Special Forces Fast Transit Craft, Military High Speed RIBs, Fast Patrol Vessels, Pilot Boats, Workboats and Passenger Ferries. He leads in the development of new products within the BMT portfolio and provides oversight and support to production detailed design teams. He has worked closely with the UK MoD in numerous roles / projects and is able to apply this knowledge in both the commercial shipping and defence sectors.

Combined academic and practical experience gained working closely with vessel operators together with a personal interest in maritime activities, provides an understanding to the practical / functional aspects of boat operations, which during requirements development aid in breaking down the operation and setting measurable performance criteria which have a direct influence on the operational characteristics of a craft.

Steven graduated in Naval Architecture at Southampton Institute, has a diploma in small craft surveying, is a member of RINA and is an Incorporated Engineer of the Engineering Council UK.

Gareth Phillips is an experienced Project Management Professional with a BSc in Mechanical Engineering from University of Wales, Cardiff. His current role as Project Manager requires the detailed development and successful delivery of a programme of marine based projects across the business, this includes the capital replacement of the Port Authorities fleet of Pilot Vessels. He has been with the Port Authority for 3+ years and brings a wealth of project management / development experience from his background in large scale Oil & Gas projects across the world. He supports the Port Authority's operational strategy and provides key direction to ensure current operations & projects are aligned with the wider organisations' key values of Safety, Excellence, Collaboration and Sustainability.

Colin Cain is the Technical Director of the BMT commercial vessel design team based in Southampton. He is a chartered naval architect with a degree and PhD from the University of Newcastle upon Tyne. His career began at the RNLI as part of the all-weather lifeboat design team, and this was followed by a period designing a wide range of commercial and military craft at VT Halmatic and BAE Systems. For the last 10 years he has led and managed a series of high performing multi-disciplinary design and consultancy teams at Longitude Engineering, Wartsila and BMT. His primary areas of technical interest are the design of small to medium sized commercial and military craft operating in onerous environments, vessel procurement, inter-operability between boats and ships, military landing craft and design for safety. His career for over 30 years in the marine sector has combined academia, research, design from concept to production, vessel trials, business development, management and leadership.

UNDERSTANDING AND MITIGATING SHOCK AND VIBRATION IN FLEXIBLE INSHORE LIFEBOATS

S Webber, P J Sheppard, T Coe and **H J Phillips** Royal National Lifeboat Institution, UK

SUMMARY

This paper presents the results of a multi-year programme of work to better understand the motions of the Royal National Lifeboat Institution (RNLI)'s D Class Inshore Lifeboat and mitigate the effect of these motions on lifeboat crew. The D Class is the RNLI's most numerous asset and as a high performance inflatable craft with the potential to generate high levels of shock and vibration. Sea trials demonstrated however, that these impacts were fundamentally different from those experienced on other vessels, with high levels of shock and vibration experienced at the stern of the vessel. The results of full scale trials both at sea and in a towing tank environment are presented to investigate the effect of vessel flexibility on motions. The paper then presents the testing of novel shock mitigation solutions to reduce the effect of impact on the craft occupants. The results of tank testing, instrumented sea trials and operational trials are presented.

NOMENCLATURE

$A_{1/100}$	Average magnitude of the top 1% of wave impacts according to standard G
$A_{1/10}$	Average magnitude of the top 10% of wave impacts according to standard G
$A_{1/3}$	Average magnitude of the top third of wave impacts according to standard G
ALARP	As Low As Reasonably Practicable
H_z	Hertz, cycles per second
HSE	Health and Safety Executive
IB1	Inshore Boat 1
LCG	Longitudinal Centre of Gravity
MEMS	Micro Electro Mechanical System
SEAT	Seat Effective Amplitude Transmissibility, ratio of vibration above and below an isolator
VDV	Vibration Dose Value given in $m/s^{1.75}$
WBV	Whole Body Vibration
W_k	Band pass filter, or weighting curve defined in ISO 2631-1 for seated individuals.

1. INTRODUCTION

The RNLI operates a fleet of over 400 lifeboats around the coast of the UK and Ireland as well as on some inshore waters. These range from large all-weather lifeboats, which are designed to operate in all weather conditions, to inshore lifeboats designed to operate closer to the shore, including in shallow water and surf.

The D-Class is the most numerous boat in the fleet, there are in excess of 180 in the RNLI including relief and training boats and are operated from over 140 stations. Powered by an inversion proofed 50hp outboard they are capable of 25kn with the crew able to re-right the boat and restart the engine if capsize should occur. They are highly manoeuvrable and usually operate closer to shore than the RNLI's larger all-weather lifeboats. Usually trailer launched from a beach they are especially useful for searches and rescues in the surf, shallow water, and confined locations - often close to cliffs, among rocks and even inside caves.

2. THE PROBLEM

The RNLI's key aim is to keep crews safe whilst rescuing others. The RNLI logs safety incidents and near misses associated with all activities, and these indicate that the D-class is at greater risk when it comes to the ride quality, slam impacts and discomfort involved in operating a D-class. As the most numerous class of lifeboat with the highest number of launches it is the natural focus for improvements to safety.

The first D-class entered service in 1963, originally constructed from a tough nylon material, water-proofed with neoprene. They could be driven at well over 20 knots via the 40hp outboard mounted on a wooden transom [1]. The boats received numerous upgrades until 1998 when a more complete review was undertaken. This review culminated in a new D-class variant, with the 5m long IB1 entering service in 2003 [2] This had composite floor and transom and a new improved pod and console forward for electronics and storage along with improvements to handholds. The composite floor retains an inflatable keel and supports a padded mattress forming the crew's deck. These changes were required as an increase in power from 40hp to 50 was also being made to increase speed, which would increase loadings on the structure and, as is now apparent, crew exposure levels.

It has long been known that the D-class presents unusual hydrodynamic challenges because they are very flexible, with the sponsons and fabric hull bottom deflect and compress at speed [3] making accurate computer or scale modelling very difficult. In addition, the boats are constructed by hand meaning each has its own minor differences. The solid scale, or segmented models as usually used in a towing tank would not reflect the flexibility of the full-scale boat, and as the fabrics used in construction are not scalable, flexible models would be fraught with difficulty. Therefore, results would be of limited benefit in understanding the boat motions and behaviours. For accurate computer modelling categorisation of the build fabrics would be required and the model validated somehow.

3. D CLASS MOTIONS

3.1 SEA TRIALS

Previous attempts had been made to understand the D-class behaviours at sea where results suggested they could exhibit unusual and high acceleration on the deck. The high accelerations increase a risk of potential crew injuries and can be in the region of 9-10g with Vibration Dose Value (VDV) of 281(m/s$^{1.75}$). However, it was not clear if these are true values of crew exposure because the composite floor is locally damped by the weight of the crew and they adapt their position to ballast and stabilise the boat in a seaway, thus moving the damping effect they have locally. The lack of understanding was compounded by the lack of any repeatability in the trials, even from one run to the next, so any validation was proving impossible. A boat was instrumented using a Fylde data logger recording at 2000Hz from MEMs triaxial accelerometers with the aim of establishing some patterns in behaviour.

When processed, the trials data appeared 'over filtered' at a commonly (for rigid craft) used low pass of 10 and 20Hz., however when using a 30Hz low pass the data appears to be showing a sufficiently good match to the peaks. This suggests that there is a significant quantity of energy within the 10-30Hz band, atypical for a rigid vessel. A 40Hz low pass was also processed but did not add any additional clarity. Whilst the need for, and level of filtering has been discussed at length elsewhere, the RNLI filters data to levels generally considered to be relevant to the human body or structure being assessed, this has the effect of reducing peaks in the data, 'over filtering' would reduce this to unrealistically low levels.

The D-class appeared to have little or no pitching prior to wave impact, also unusual is the negative acceleration after the impact. In larger craft, the hull sinks into the water upon impact, which maintains a positive "hump-phase" acceleration after the impact. Here the hull does not sink much during impact so there is an immediate feeling after impact of a semi-free fall phase figure 1 below.

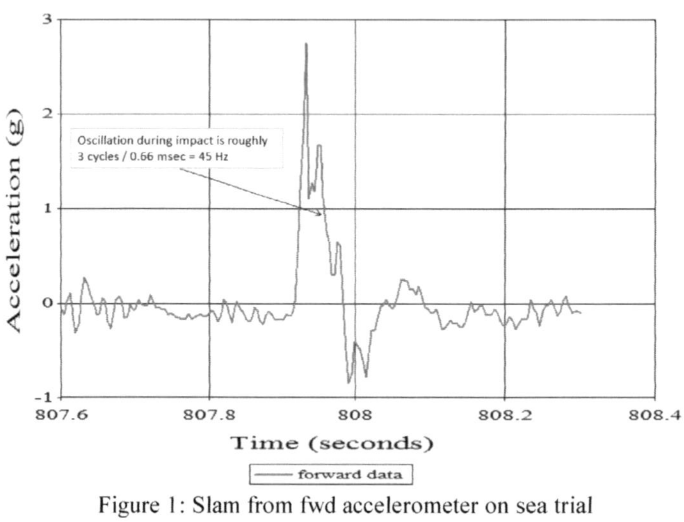

Figure 1: Slam from fwd accelerometer on sea trial

The results demonstrated further anomalies; in some trials the composite floor near the transom saw higher accelerations recorded than those in the bow. This appeared regularly but appeared not to be predictable, in addition the forward and aft peaks appear to be consistently out of phase. The example here shows slams where the bow, as would be more usual, accelerates more than aft, however other impacts reverse this, but the out of phase response remains. In addition, there appeared to be a whipping motion along the length of the hull, with hints of this also detectable in the slowed down video footage where the middle of the boat could be seen 'bouncing' but required further investigation.

4. TANK TESTING

4.1 ORIGINAL TOWING TANK WORK

With the sea trials providing no clarity it was decided to revisit the 2003 towing tank work where despite the difficulties above, a solid model was made, tank tested and compared to the resistance, propulsion, LCG variation, trim, and sinkage of those of an existing earlier full-sized version of the D-class. This was in readiness for the design changes to the new IB1 variant where, amongst other things, the wooden floor and transom would be replaced with composite components and engine size would be increased from 40 to 50hp. To investigate the accuracy of the model against the actual boat both were run in the towing tank operated by QinetiQ in Gosport UK.

Despite the 270m(L) x 12.2m(W) x 5.4m(D) dimensions of the tank it was not of optimal size for the 5m length and 2m beam of the full-size D-class but did offer controlled, repeatable conditions and enabled full instrumentation of the boat and a theoretical speed of 23.7kn could be achieved [4].

By comparing the solid model with a flexible boat, various observations could be made. Taking resistance first, this arises from the lift force obtained from the large positive pressure peak abaft the spray root, resolved into the direction of motion. As this depends on running trim angle, it seems at first counter-intuitive that the model, running at the higher trim angles, has the lower resistance in the planing regime. However, the increase in bow-up trim reduces the wetted area, thereby reducing the viscous resistance; this perhaps accounts for some of the lower resistance predicted by the model. However, the "residuary" resistance arises from lift forces resolved over the hull surface in the direction of motion so, if this surface is distorted, especially near its forward end with its greater deadrise, greater resistance will accrue from the less streamlined, distorted, hull shape. Thus, it would seem that hull distortion of the full-scale, "flexible" craft nullifies any benefit from its reduced running trim.

Mean sinkage/rise comes from the vertical resolution of the normal pressure force and is dominated by the magnitude of the heave force derived from the pressure distribution over the bottom, not the form of the distribution itself. The same cannot be said for trim which is determined by the moment of the heave pressure force distribution about, say, the LCG, and is thus dependant on the "shape" of this distribution. A reduction in heave pressures over the forward end of the planing surface, due to distortion, would lead to a reduction in bow-up trim moment leading to the reduction in bow-up trim angle measured on the full-scale, flexible, craft.

It is perhaps intuitive that trim is sensitive to hull shape and the "shape" of the distribution of normal pressures fore and aft, whereas mean sinkage/rise is not. This perhaps explains why mean sinkage/rise is well predicted by the rigid model because, as its mass is scaled correctly, the resultant heave force or "lift" when planing (which must support most of that mass because hydrostatic support is low) is also scaled correctly and therefore equivalent at both model and full scale. Therefore, it might be expected that the sinkage/rise, which depends on lift, to be well-predicted by the rigid model, this is indeed the case, but such is not the case with resistance and trim. These findings may be related to the origins of resistance, sinkage, and trim as all of these effects are derived from hydrodynamic pressure forces normal to the hull surface, resolved in the appropriate direction. This makes the reasonable assumption that, in planing conditions, normal pressure forces dominate "lift" and forces tangential to the surface dominate viscous resistance. The "residuary" components of resistance, mean sinkage and running trim when planing, all result from normal forces resolved over the hull surface, whereas viscous ("friction") forces derive from the tangential components over the wetted area of the hull.

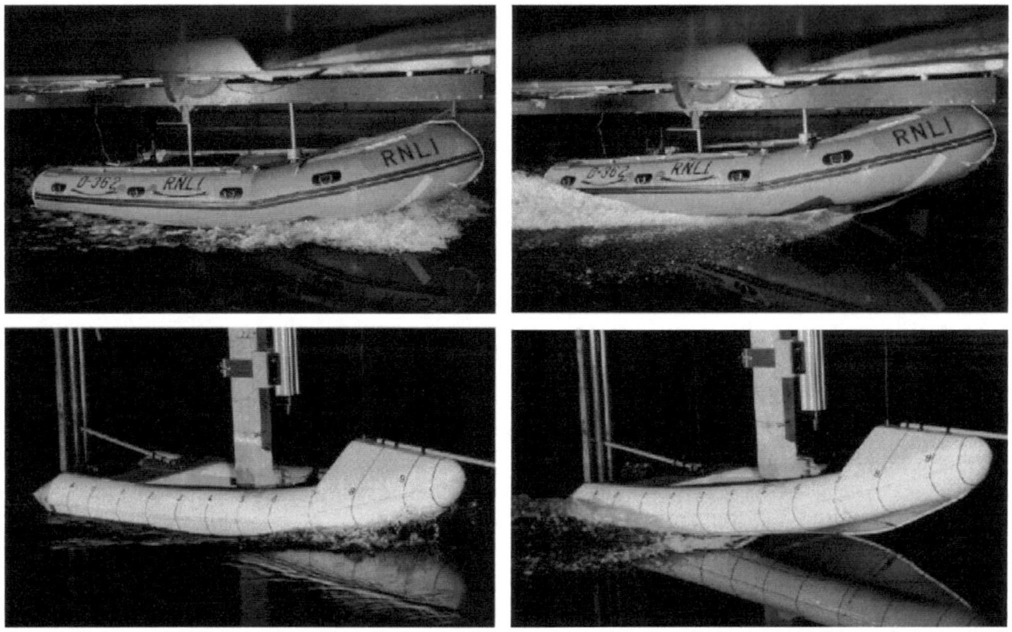

Figure 2: D-362 and associated model at 5.8 (left hand images) and 19.4 knots (right hand images).[5]

Figure 2 shows a comparison of the D-class D-362 and model at two speeds. The lower speed is 5.8 knots, well before planing has commenced, and the higher is 19.4 knots, well into the planing regime.

From this Figure the following observations may be made:

- At the lower speed, the bow breaker and wave emanating from the forward vertical knuckle around station 6 are similar at both model and full-scale, although the spray is more fragmented at full-scale possibly as the boat is not a fair as the model. At this speed, resistance, mean sinkage/rise, and trim of the rigid model are close to those of the full-scale craft. See figure 3.
- At the higher speed, similar wave and spray formation may be noted and the increased bow-up trim of the rigid model is clear. There was significant air entrainment in the full-scale craft, in the form of vortices under the hull, apparently emanating from the sponson/bottom junction at the spray root. It is not known if entrainment occurred with the model, but the wave/spray formation seems to be similar at both model and full scale.

The model's running trim is noticeably greater than that of its full-scale counterpart because it is running on the pressure peak just abaft the spray line; at full scale, bottom distortion means the pressure peak does not have so great an effect so trim is reduced and resistance increased.

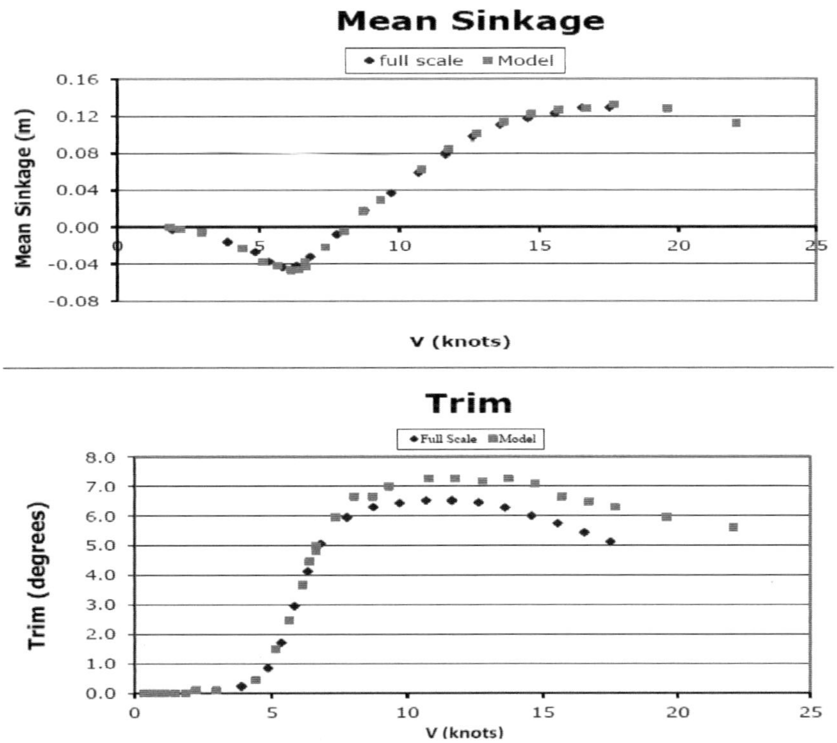

Figure 3: Sinkage and Trim of D-362 and that predicted from its model.

All these trials were completed in calm water, so clarified some of the boats' behaviours, demonstrating the extent of bottom panel deflection for instance and also showing that for some criteria a solid model could act as a predictor of a flexible boat's behaviour.

Further trials were then made with the model into head seas, to investigate the question of whether resonance was being caused by impact with the water surface, during which the natural pitch and heave periods of the D-Class model were measured. The natural pitch frequency of the D-Class model was measured as 1.4 Hz which may be compared with the frequencies for which the secondary pitch (and heave) response occurred. These ranged from 0.9 Hz to 1.5 Hz with a mean of 1.25 Hz. This is sufficiently close to the natural pitch frequency to confirm that resonant motions were, in fact, being triggered in both regular and irregular waves [6].

The model was also seen to experience non-linear behaviour, especially at higher speeds, but despite this, a conventional transfer function (or its square, the response amplitude operator) was calculated, for regular and irregular waves. These values compared well, but only at low frequencies. It was concluded that this method may have some value despite the non-linear responses but would need validating in full scale D-class trials to ascertain the effects of using a rigid model to predict the seakeeping behaviours of a flexible, inflatable boat.

4.2 DROP TESTING

From a better understanding of whole-body vibration, improved reporting systems and a review of the D-class, further understanding of the D-class motions was required such that further improvements to crew safety could be investigated. Deflection was evident in the fabric bottom from the earlier towing trials and was expected to play a significant effect in the slamming accelerations, but the extent of its effect and behaviour were not known. Drop tests were planned to try and understand the simpler dynamics without forward motion, the RNLI's sea survival training pool was used to conduct these instrumented trials. This approach had previously been used with models by other operators [7] but not with full sized or flexible, fully inflatable craft.

Within the sea survival pool the D-class could be dropped into both calm water and waves. The boat was suspended using standard lifting strops that could be adjusted to correct trim as required and ballasted to represent either 2 or 3 crew before being dropped from both 0.5 and 1m. The boat was released remotely using a line to a slip hook and was unguided and unrestrained in free fall.

Due to the fabric hull sections and the likelihood of this impacting the internal composite floor and the plan for further trials it was thought impractical to fit pressure sensors to the hull skin as had been used on previous RNLI drop tests to study pressure distributions. A rate gyro and 3 triaxial accelerometers were fitted, and along with multiple camera angles would be analysed post trials. 15 drops were carried out into flat water with a further 6 into waves.

When the data was analysed, some unfiltered accelerations were observed to be clipping at 45-50g in both positive and negative directions. This would lead to some error in calculations where very high VDV levels were seen. It is acknowledged that VDV may not be the most useful metric in this situation but allowed some correlation with sea trial results and data from other RNLI classes during whole body vibration trials.

Forward and aft impacts were seen to be more severe than at midships in all but one drop. The reason for this anomaly remains unknown but it was a drop in calm water so was not the result of the wave pattern in the pool. The midships impacts were generally more confused and very slightly delayed from the forward and aft impacts. This is possibly one of the many errors associated with drop testing small flexible boats. Indeed, all results should be treated with caution from this drop trial study as it had not been possible to consider or calculate all possible error margins or validate results.

When the slow-motion footage was studied it became apparent that the entire boat was considerably more flexible than had been envisaged or was apparent from sea trials, with the energy from the drop reverberating, especially around the tubes. As per the sea trials this appeared to be driven by the weight of the forward pod.

4.3 FURTHER TOWING TANK WORK

All the original towing work and the majority of the drop testing was carried out in calm water. This, coupled with the drop testing results and the non-repeatability of sea trials resulted in a decision to carry out some further towing tank trials, this time in waves. The object was not to repeat the 2003 trials but to instrument the D-class to try and understand the dynamic motions in regular wave patterns at different speeds. With modern modelling techniques it was theoretically possible to create a computer model of the D-class to analyse its seakeeping performance but experience in other projects had shown this to be complex and costly, so this option was discounted.

Since the resistance and other seakeeping measures were not to be included in these trials a traditional towing carriage setup was not required. Instead, the aim was to tow the boat in such a way as to simulate as closely as possible the propulsion from engine thrust but due to concerns about water contamination an engine or even a blank engine leg could not be fitted. There also had to be some compromise to ensure the boat remained on a forward heading, and safety lines were rigged in case of failure.

Data collection details were identical to those fitted for the drop tests with the addition of some visual gauges fitted through the composite floor to the fabric bottom so the footage from onboard cameras could be used to measure the deflections of the fabric hull.

A total of 40 runs were made with variations made to keel pressure, carriage speed and wave frequency. Keel pressures used were 3psi and 6psi, speeds were 4 to 12m/s (Approx 7.5 to 23kn). The limits of using the tank meant that the wave heights had to be kept small at 0.1m. During sea trials at speed in 'calm' water this wave height was known to cause significant crew discomfort and for the high-speed runs, the wave height had to be reduced to 0.075m but was still sufficient to cause significant boat motions. Regular wave patterns were used throughout so as to limit the variables. Ballasting for the 3rd crew member caused a slight list and without the engine leg to aid directional stability this caused the boat to veer off course so only 2 crew members could be ballasted for.

Significant datasets were collated with unknown distributions, so a Spearman's correlation was used, this demonstrated that only the wave encounter frequency appeared significant. Boat speed and wave amplitude for example appeared less important overall indicating that the situation is likely to be complicated by other parameters having different effects but there is currently not enough data to comment on what these might be. Keel pressure increases showed little effect on results demonstrating it was unlikely that this could be used to minimise the deflections seen in the hull bottom fabric.

From studying the onboard videos, the console forward appears to move with a higher frequency than the encounter frequency, with the forward part of the boat bending on impact. At higher encounter frequencies the console passes through a mode where it vibrates at twice the encounter frequency, but this was not clear on all higher encounter frequency runs. On some runs at very similar encounter frequencies the console was sometimes in phase and sometimes at a higher frequency than the encounter frequency, even when in phase a higher amplitude than that of the underlying boat was observed. The worst boat encounter frequencies forward are 3 to 3.5Hz where both boat bending and console movement relative to the boat are both involved. Midships where the 'bouncing' had been observed in the sea trials footage at a 2.5Hz encounter frequency resulted in the higher accelerations. It was not possible to confirm if these 'resonances' are true resonance of a natural vibration or are a purely wave driven forced effect. These are above the 1.25hz measured on the head seas model trials.

Spectral analysis did not reveal any obvious boat resonance. This is possibly due to the heavy items of console and ballast weights in a flexible boat causing several resonances emanating from the flexible couplings, different driving forces and changing force directions. If the boat was sitting on a shaker platform it may be possible to identify the various hull and equipment resonances one by one, but the current sea and tank data is too complex to evaluate for this purpose. The power spectral densities were in part helpful in explaining the changes in VDV with encounter frequency.

Some test runs and locations resulted in energy mainly at one frequency. There is one large peak during a 10m/s run with a relatively high 4.4Hz encounter frequency and then the energy falls away with increasing frequency, suggesting a sinusoidal motion for this location and test condition. This large peak falls into the maximum pass band of $W_{k.}$ weighting, from around 4 to 8 Hz, so the human body will experience this motion. Other test runs and locations however showed several strong harmonics at multiples of the wave encounter frequency, suggesting a more distorted, less sinusoidal motion.

The 7 m/s dataset is similar with respect to the relationship between change in VDV and amount of energy in the $W_{k.}$ passband. The rise and then fall in VDV at this speed in the forward location with an increased encounter frequency is explained (in part) by the increase and then decay in energy at frequencies at a multiple of the encounter frequency, as well as the increase in the frequency of the (fundamental) wave encounter frequency, both of which alter the energy in the $W_{k.}$ passband.

Interestingly, the increase in the relative amount of energy at the first harmonic of the encounter frequency in the midships and aft locations at higher encounter frequencies in the 7m/s data does not show in the 10 m/s runs for the equivalent encounter frequencies. It is perhaps not surprising that the boat is responding differently when encountering more closely spaced waves at a lower speed compared to a condition with more widely spaced waves encountered at a somewhat higher speed. The data here however gives an example where two very similar, but not quite identical, encounter conditions result in quite different behaviour.

It has been shown then that an inflatable D-class has complicated, non-linear responses that vary across different areas of the boat, and that in benign conditions is still capable of imparting substantial shocks to the crew. Furthermore, there was little in the current boat that could be used to mitigate the crew shocks, so other shock mitigating solutions would be required.

5. NEW MATRESS SHOCK MITIGATION TRIALS

As discussed above crew are known to experience discomfort as a result of the exposure levels experienced in the boat. To mitigate the effect of these impacts an initial review by operators and engineering staff was carried out using the standard HSE Hierarchy of Controls [8] and considering the vessel overall. Whilst other options were identified, it was decided that improving the isolation of the crew from the composite deck was the preferred option to develop.

5.1 D-CLASS SHOCK MITIGATION OPTIONS

To manage the risk of being thrown overboard, crew in the D-Class are required to kneel on a foam pad attached to the deck of the boat. This pad is normally referred to as the mattress and is made of foam selected to be statically comfortable and as such does not offer much in the way of shock mitigation in a dynamic environment.

Alternatives were reviewed and four were shortlisted for trials, two were inflatable and two were foam-like materials. The nature of D-Class operations means the risk of a deflated mattress is deemed intolerable due to the risk of injury, so despite air springs offering benefits, the two inflatable options were not progressed on safety grounds.

Other possibilities for improving and controlling hydrodynamics to reduce the harsh boat motions were considered and are still being trialled but mattress options were assessed as being the quickest way to improve crew safety and comfort whilst also being a practicable solution. Reducing D-Class WBV exposure requires a holistic approach where benefits from multiple engineering solutions are applied cumulatively.

5.2 INSTRUMENTED TRIALS

To test the manufactures claimed shock mitigation performance of the two selected mattresses a series of instrumented trials were run from The RNLI Lifeboat College in Poole.

5.2.1 Methodology

Classic methods of measuring vibration exposure and shock mitigation [9] do not lend themselves to the quantification of the vibration exposure when kneeling. The need for the accelerometer to be coupled to the human body under investigation is usually achieved with a seat pad. This is not practicable with a kneeling posture, furthermore if a person leaves the deck and lands back on the sensor large transients are recorded which are not representative of the actual vibration environment.

To address these issues and provide an estimate of the vibration mitigation performance of the two candidate mattresses, body mounted accelerometers were used. Gulf Coast Data Concepts Human Activity Monitors (HAM) loggers were fitted to TENS machine medical grade pads and mounted to the side of test subjects' knees. Vibration levels were then compared with those measured on the deck of the boat using two Gulf Coast Data Concepts HD 50 accelerometers (see figure 4).

A series of trials on three different mattress types were undertaken:

1. Competitor A Layup I
2. Competitor A Layup II
3. Competitor B

Each run consisted of three repetitions of the same track with the crew rotating between positions to achieve an average acceleration dose over the boat.

- Post processing was undertaken using MATLAB 2019b and calibrated using the calibration rates provided and confirmed via freefall and rollover checks.
- Deck data was filtered with a 4th order Low Pass Butterworth Filter with a cut off frequency of 20 Hz and visually inspected for data acquisition issues, errors, or unexpected artifacts.
- VDV and standard measurements were calculated in accordance with ISO2631-1 [10] and Riley et al [11].

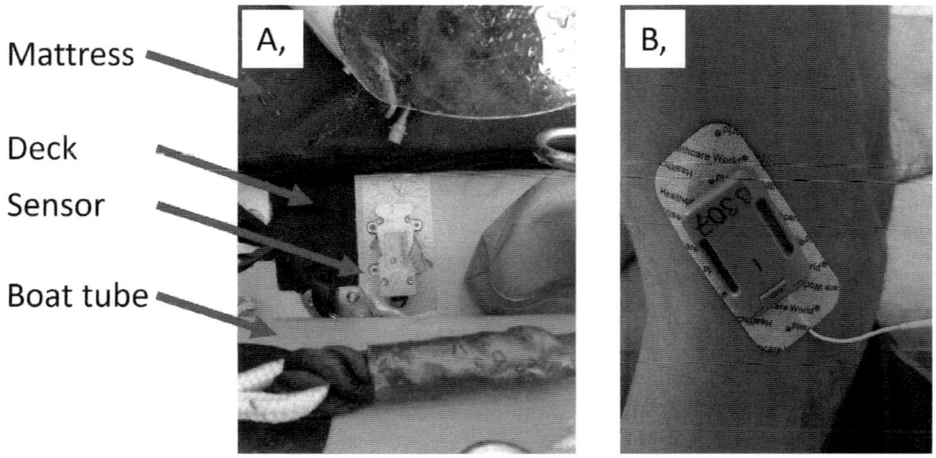

Figure 4: Trials sensor setup. A) Deck mounted triaxial accelerometer mounted in line with crew position B) Knee mounted accelerometer, strap omitted for clarity

5.2.2 Results

Figure 5 shows all three runs while comparing deck accelerations in Table 1 show that all three runs were undertaken in similar conditions.
Each run lasted approximately 50 minutes, made up of 3 repetitions of the same 15-minute trials pattern with the additional 5 minutes made up of transit times and crew rotations.

Deck acceleration measurements were used to assess the vibration environment and are typical of operating the D Class in short chop, with VDVs of 56, 43 and 47m/s$^{1.75}$ respectively for each transit. Due to the lack of mass associated with the accelerometer attachment points the VDV will be an overestimation however it is appropriate for comparative purposes. Comparing VDVs and standard G [11] values shows that the first run was most severe but all transits were extreme shock environments.

To assess mattress performance the vector sum of the on knee acceleration was calculated. Using the vector sum accounts for changes in posture and movement within the boat. The VDV was calculated using the W_k filter defined in ISO2631-1 [10] because whilst this is not intended for vibration entering at the knee it was felt it was more appropriate than a weighting filter defined for standing personnel. To account for inter-subject variability the average acceleration of two crew was used and compared with the deck accelerations.

As table 1 shows, Competitor A performed significantly better than Competitor B and in its second configuration shows 24% shock mitigation, comparable to the manufacturers claimed 30% reduction in laboratory conditions.

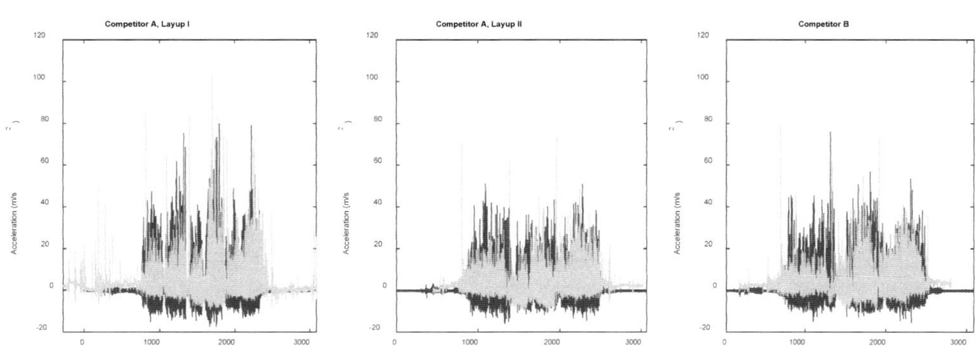

Figure 5: Acceleration comparisons for three trials mattress serials showing deck acceleration (blue) and knee acceleration (green)

Table 1. Deck vibration for runs 1, 2 and 3 showing VDV, time to limit value and Standard G metrics as well as mattress performance.

		Competitor A layup I	**Competitor A layup II**	**Competitor B**
Duration	(hh:mm:ss)	00:50:01	00:51:07	00:51:26
VDVx	($m/s^{1.75}$)	8.24	6.89	6.94
VDVy	($m/s^{1.75}$)	9.72	8.14	8.66
VDVz	($m/s^{1.75}$)	56.16	43.55	46.52
Time to 9.1 VDV	(hh:mm:ss)	00:00:02	00:00:06	00:00:05
Time to 21 VDV	(hh:mm:ss)	00:00:59	00:02:46	00:02:08
$A_{1/100}$	(g)	3.83	2.87	3.17
$A_{1/10}$	(g)	2.47	2.01	2.21
$A_{1/3}$	(g)	1.76	1.43	1.57
Knee 1	($m/s^{1.75}$)	56.61	38.43	41.06
Knee 2	($m/s^{1.75}$)	42.19	28.04	50.37
Mean	($m/s^{1.75}$)	49.4	33.24	45.72
SEAT		**0.88**	**0.76**	**0.98**

5.3 COASTAL TRIALS

In order to determine whether the measured reduction in vibration translated into a reduction in the reported discomfort when using the D Class, a double 'Before and After Controlled Intervention' study was developed using five prototype mattresses based on the design of Layup B above. Ten stations were identified in two groups that were selected to be representative of the complete operating environment the vessel. Two stations were selected for each of 5 key groups:

- Surf, producing some of the largest impacts in breaking waves.
- Busy English Channel stations, with a high number of launches.
- Short fetch, short steep waves produce unpleasant impacts even when wave amplitude is low.
- Estuary, stations often using their vessel in sheltered waters, both on the east coast of the UK.
- West facing, identified as launching in bigger than average sea states, often with long wave lengths. The antithesis of the short fetch stations.

5.3.1 Data collection

Crew had the opportunity to record any discomfort or issues using an in house questionnaire implemented in Qualtrics (www.Qualtrics.com). Operators recorded mission duration, sea conditions and location within the boat as well as instances of "discomfort" using a modified Nordic Pain Questionnaire [12].

5.3.2 Results

Across both groups 479 individual reports were made, split between control (215 responses) and intervention (269 responses) mattress. Crew were asked to record their position in the boat which demonstrated that for both the control and intervention conditions the expected range of positions was obtained. Approximately 1/3 of all entries were by people helming the vessel and crew were distributed between the two normal crew positions. As the vessel is often reported to be more comfortable at the stern it was important to ensure the data set was not skewed either way.

Figure 6 shows the response to the overall comfort level. This shows more than 50% reduction (112 reports to 48) in discomfort and a corresponding increase in comfort. The mattress also exhibited a polarising effect where there is a reduction in people reporting neither comfort nor discomfort, with the bulk of these responses moving to 'comfortable'.

As well as overall comfort the survey invited respondents to report discomfort in a particular area of the body. An example of these results is shown in figure 7 below. The data agree with anecdotal reports that kneeling causes particular discomfort to the knees as well as the more typical lumbar spine discomfort reported in small boats with seated occupants. The number of reported lower back discomfort incidents is dramatically lower with the intervention mattress. The number of people reporting any discomfort at the knee fell slightly. The effect was most prevalent in surf stations where high levels of slamming is typical but was also seen in all trials stations and in all sea conditions.

The reduction was largest in short voyages of less than one hour, with the percentage improvement over foam reduced for long duration voyages. This is thought to be as discomfort from prolonged periods kneeling increases in both cases regardless of mattress type.

Some stations in calm conditions reported that the mattress was initially harder than their previous mattress. The supporting data however still showed an improvement in terms of reduced instances of pain and soreness in these conditions.

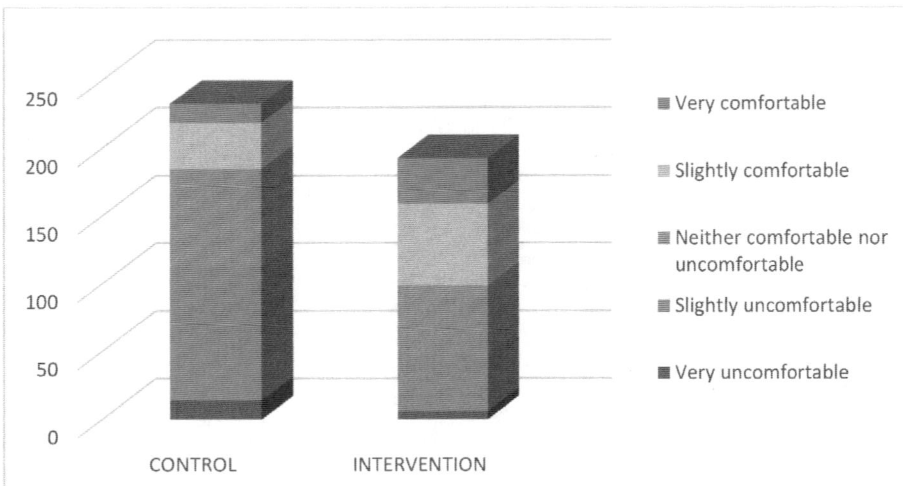

Figure 6: Overall comfort level reported with Control and Intervention

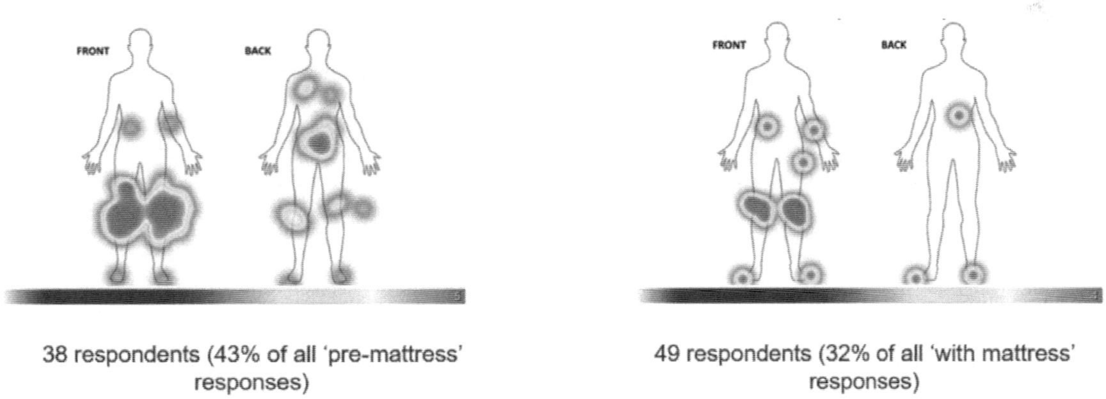

Figure 7: Group A number and location of reported discomfort while afloat **(Note: the heat map shows number of reports not intensity)**

6. CREW FEEDBACK

Shock mitigation in a scientific form does not necessarily translate into a usable piece of equipment. During the trial, crews were asked to provide feedback on the intervention and control mattresses to gauge opinion on the mattress change from a subjective comfort and usability perspective.

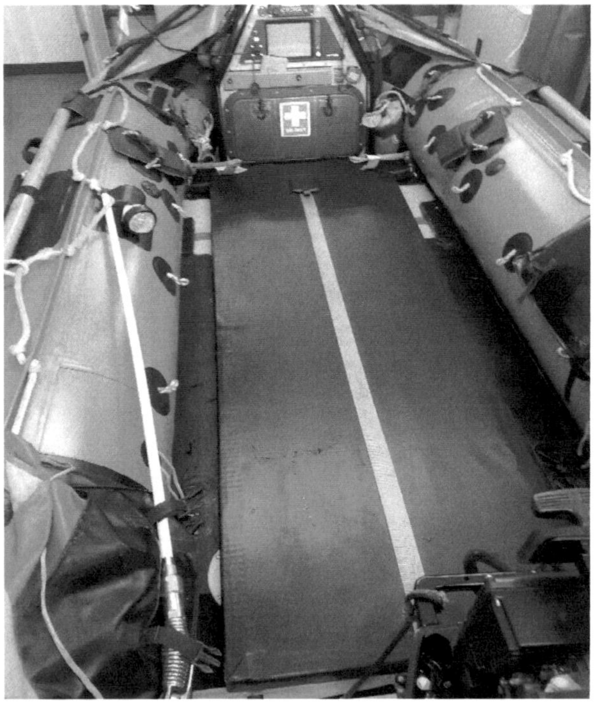

Figure 8: Trial mattress in situ.

The trial feedback on the intervention mattress included concerns over water ingress through the cover, a lack of grip when wet, firmness on knees, firmness in light conditions and restricted access to key boat safety features, such as the keel inflation valve.

The feedback was predominantly positive with regard to the intervention mattress responsiveness in waves but its firm nature in more benign conditions was noted by many.

The trials feedback on the existing foam mattress focuses on its improved grip and short-term comfort but highlighted the increasing discomfort in legs, knees, back and posture as sea state increases.

The feedback from the trial has shown the intervention (see figure 8) is a positive addition providing an improved cover can be developed.

It is acknowledged that this is an improvement not a cure to the potential discomfort of operating this class of boat. By the very nature of the kneeling position in a D-Class it would be unlikely to find any solution that eliminated the discomfort experienced from long term adoption of this posture.

7. PRACTICAL IMPLEMENTATION OF DESIGN IMPROVEMENTS

The most significant challenge in the mattress improvement was finding a suitable cover which met the RNLI requirements and allowed the intervention to perform effectively. A watertight, retrofittable solution which gives crews at least the same level of grip and fitout as the existing mattress was required to gain approval that it was safe to operate. In addition, the intervention material was 'tuned' by trialling different combinations of products available to suit both the collated data and subjective feedback.

Mattress design iterations are shown in Table 1 and summarise the efforts taken to meet operational requirements whilst preserving the effectiveness of the intervention core. Initially in-house skill sets were utilised; however hand gluing was not producing the robustness or watertightness required and was quickly eliminated in favour of Radio Frequency (RF) welding. RF welding is a process which uses high frequency electric fields to induce heating and melting of thermoplastic based materials. The advantage of this process is that joints and seams in the cover are fast, permanent, repeatable, tidy and can be tooled to reduce the influence of operator skill on quality and repeatability. The move from Hypalon® rubber to Polyurethane fabrics for the cover material was largely driven by fabric properties, specifically fabric elasticity and drapability. The material selection also aided the change from hand gluing to Radio Frequency (RF) welding in later iterations of the design.

Table 2. Overview of mattress design iterations

Version	Description	Image	Pros	Cons
1	Zipped Sail bag cover	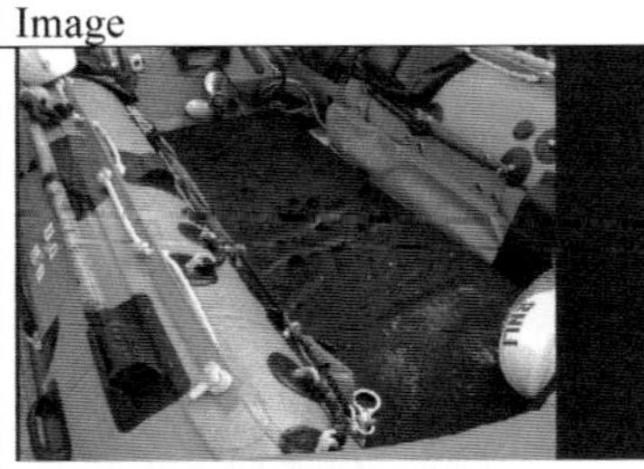	Quick solution for intervention core proving trials. Enabled simple swapping of intervention Core layers.	-Lacking important safety features - centreline strap and securing points -Not watertight, -No access to inflation valve hole
2	Hand cut, glued PU fabric cover	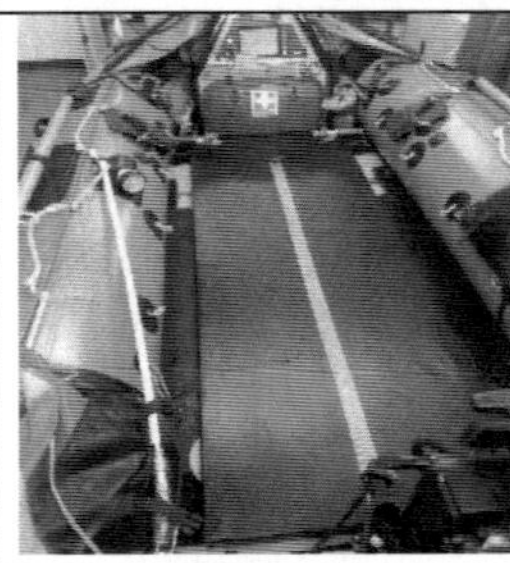	Used in house materials, experience, and labour Inexpensive option for intervention core proving trials.	Highly labour intensive Not robust - proved to degrade/fray quicky during trial, Is not easy to productionise - high labour cost and manufacturing time
3	RF welded two part prototype		Robust low labour solution, watertight, allows multiple fabrics to be used in construction	External manufacturer required, Issues with capsize safety due to fixing straps
4	RF welded single unit		Improved securing point to maintain air gap in the event of capsize Removal of Velcro joining pad for improved long-term robustness Increased thickness of foam "ears" to protect from deck equipment.	Production time high as still using prototype techniques not tooling. Final shaping still to be confirmed by operations team. Some minor issues regarding watertightness and final fit out to be addressed. Foam topper required for calm condition comfort.

- Final Production
  - Tooled production to streamline output from small supplier.
  - Initial rollout to 10 stations to validate robustness to chemical attack & crew usage.
  - Full fleet wide roll out planned during 2024 season (140 boats in total).
  - Returned mattress disposal strategy in place in accordance with RNLI sustainability program.
  - Repair / replacement cover strategy in place for ongoing through life support

8. CONCLUSIONS

Undertaking both drop testing and tow tank testing gave valuable insights into how the D-class behaves at sea that was not possible to learn during sea trials. The tow tank trials of a flexible full-sized boat compared against a rigid scale model also demonstrated that whilst not all data could be carried across, a solid model could provide accurate predictions of some design criteria. Unfortunately, at the time of the original IB1 development this model validation to full size was not fully explored which led to a long delay in better understanding the seakeeping characteristics of the D-class. In addition, if the inflatable boat could be made to behave closer to the solid model, then there would be potential benefits for handling and shock reduction; This is part of an ongoing investigation for future D-class variants.

Measurements taken with body worn accelerometers have been shown to give useful results for comparative purposes, although it is not possible to implement wave by wave methods in the analysis of the data. In addition, care is needed as to where on the body accelerometers may be positioned to minimise effects of the individual's physique as well as consideration of the measurements to be made, and the safety of the wearer.

The reductions in discomfort are substantial and correlate well with VDV SEAT metric reductions in the specific context of kneeling persons in a small boat. This is a significant conclusion demonstrating the effectiveness of the intervention and providing some elements of validation of the VDV for shock levels above that for which it was developed.

Finding an effective solution that can be productionised and balances the competing D-Class requirements in service has been challenging with no single ultimate solution having been found to an extremely complex crew comfort challenge in a boat with complex motions. However, the RNLI believes the introduction of a shock absorbing cored mattress is a step progression of D-Class shock mitigation from a foam mattress and goes some way to improve crew safety and to becoming ALARP within the constraints of an in-service RNLI asset. Other longer-term options for future developments and variants are still being investigated which will also include a re-evaluation of the requirements of the boat as a whole to ensure the risks to crews are minimised as much as possible.

9. ACKNOWLEDGEMENTS

Andrew Clarke – for undertaking the main trials.
Ian Dand – for technical knowledge and guidance.
Mike Riley & Tom Gunston – sea-trial analysis and insights
QinetiQ – for allowing use of their tank in unusual configurations
Sabine Courtice – for implementing the discomfort questionnaire in Qualtrics
RNLI Lifeboat trials stations volunteers – for providing trials feedback
The rest of the RNLI Whole Body Vibration, Noise and Ergonomics Project Team – for help, support and patience, with particular thanks to Claire Allard, Maddie Ball and Matt Bramley

10. REFERENCES

1. RNLI, 'The Story of The Life-Boat', 1964

2. AUSTEN, S J., and FOGARTY., 'Design and development of a new inflatable lifeboat' *SURV 6*, 2004

3. DAND, I., 'Inflatable ILB hydrodynamics and flexibility', *RNLI report*, 2016.

4. QinetiQ., 'Towing Tank', *Data Sheet*, 2013

5. BMT SeaTech, 'The effect of LCG on the hydrodynamics of an RNLI D-class inflatable' *internal report* 2003

6. BMT SeaTech, 'RNLI D-class seakeeping measurements in head seas' *internal report* 2004

7. PETERSON, WYMAN, and FRANK., 'Drop tests to support water-impact and planning boat dynamics theory' *Naval Surface Warfare Centre*, 1997

8. HSE, Using personal protective equipment (PPE) to control risks at work, https://www.hse.gov.uk/ppe/managing-risk-using-ppe.htm

9. GRIFFIN, M J., 'Handbook of Human Vibration', 2004

10. International Organisation of Standardisation, ISO 2631-1:1997 Mechanical vibration and shock — Evaluation of human exposure to whole-body vibration — Part 1: General requirements

11. RILEY, M., HAUPT, K., COATS, T., GANEY, H., and MURPHY, H., 'A Guide For Measuring, Analyzing, And Evaluating Accelerations Recorded During Seakeeping Trials of High-Speed Craft', *NSWCCD, West Bethesda*, 2016

12. CRAWFORD, J., 'The Nordic Musculoskeletal Questionnaire', *4, Oxford: Occupational Medicine*, 2007, Vol. 57.

11. AUTHORS BIOGRAPHY

Susie Webber C.Eng MRINA is a Senior Naval Architect at the RNLI. She has worked at the RNLI for over 10 years and has extensive knowledge and experience in small craft design, stability and composite boat building. She is actively involved in safety case development and equipment design for safety improvements on all RNLI assets.

Pete Sheppard C.Eng MRINA is a Senior Naval Architect at the RNLI. He joined the Royal National Lifeboat Institution in 2006 with extensive knowledge and experience in composites and boat building. He has been running a program of measurement, research and development into Whole Body Vibration and its effects across the RNLI's fleet for 12 years. He now combines this learning with the particular human factors seen on Search and Rescue vessels.

Tom Coe C.Eng PhD MRINA is a Senior Engineer at the RNLI. He joined the RNLI in 2019 as a WBV, Noise and Ergonomics specialist from the Ministry of Defence where he lead their WBV programme. He's worked in the field of WBV since 2005 and is the Chair of the BSi Panel on Human Response to Whole Body Vibration and formerly the UK representative to the ABCD working group on Human Performance at Sea. He's a volunteer RNLI crewmember and powerboat instructor.

Holly J Phillips MBE CEng PhD FRINA is the RNLI's Principal Naval Architect. She joined the RNLI in 2002 whilst also previously serving as an RNLI crewmember. She is the delegated design authority for naval architectural aspects of the RNLI's new and existing fleet and currently sits on the RINA Maritime Innovation Committee.

The Royal Institution of Naval Architects
8-9 Northumberland Street
London, WC2N 5DA
United Kingdom

ISBN 978-1-7138-9564-0

CURRAN ASSOCIATES INC.
proceedings.com